LAND AND WINE

LAND AND
WINE

THE FRENCH TERROIR

Charles Frankel

WITH A FOREWORD BY JOHN VARRIANO

THE UNIVERSITY OF CHICAGO PRESS

CHICAGO AND LONDON

The University of Chicago Press, Chicago 60637
The University of Chicago Press, Ltd., London
Originally published as *Terre de Vignes* © 2011 Editions du Seuil
English Translation © 2014 by The University of Chicago
All rights reserved.
Published 2014
Paperback edition 2021
Printed in the United States of America

30 29 28 27 26 25 24 23 22 21   1 2 3 4 5

ISBN-13: 978-0-226-01469-2 (cloth)
ISBN-13: 978-0-226-81672-2 (paper)
ISBN-13: 978-0-226-01472-2 (e-book)
DOI: https://doi.org/10.7208/chicago/9780226014722.001.0001

This work, published as part of a program providing publication
assistance, received financial support from the French Ministry of
Foreign Affairs, the Cultural Services of the French Embassy in the
United States, and FACE (French American Cultural Exchange).
French Voices logo designed by Serge Bloch.

Library of Congress Cataloging-in-Publication Data

Frankel, Charles, author.
    Land and wine : the French terroir / Charles Frankel.
    pages ; cm
    Includes bibliographical references and index.
    ISBN 978-0-226-01469-2 (hardcover : alk. paper) —ISBN
978-0-226-01472-2 (e-book) 1. Viticulture—France—Guide-
books.   2. Wine districts—France—Description and travel.
3. Grapes—Soils—France.   4. Soils—France.   5. Geology—
France.   6. Wine and wine making—France.   7. France—
Description and travel.   I. Title.
    SB387.8.F8F755 2014
    643.80944—dc23

                                                    2013027584

CONTENTS

Foreword · vii
Preface · ix

FOREWORD

Fine wine has always challenged those seeking to define its mysterious qualities. Identifying the type of grape, its regional origin, and the date of its transformation into wine has intrigued connoisseurs since the beginning of history. The earliest indication of what we now call "vintage year" and "appellation contrôlée" harks back more than three millennia to the age of Tutankhamen in Egypt. When the pharaoh's tomb was excavated in the nineteenth century, dozens of labeled wine jars were found inside; each label noted the year of Tut's reign in which the jar's contents had been made, and the location and proprietorship of the vineyard. Grape varietals were also familiar in the ancient world: Homer informs us that Laertes, father of Odysseus, planted over fifty varieties of grape in different parts of his own vineyard, while Democritus claimed to know every vine growing in Greece.

It was Pliny the Elder's *Natural History* (79–77 BC) that set the tone for all subsequent inquiries. Book 14 of his magnum opus offers a catalog of all the wine varieties known to him. Of those, he regarded Aminean, Nomentan, and Greculan as the best, although ampelographers (botanists specializing in grapevines) debate their survival into the modern age. Pliny spends more time in this work making regional distinctions, singling out some—mostly in Italy—that retain their legendary qualities to this day. Furthermore, he was aware that wines "can improve with age," an observation especially true of the Aminean grape he claimed was his very favorite.

Pliny's strongest link to modern wine studies is his observation that winemaking "be adapted to the peculiarities of the climate and the qualities of the soil." In a later passage, he recommends that certain varieties be planted in "localities fully exposed to the sun" and others in "denser soil or a locality more liable to fog." Certain hybrids, he says, "are suited to any kind of soil." Few improvements were made to Plinian classification during the Middle Ages and early Renaissance, but by the middle of the sixteenth century, Pope Paul III's *bottigliere* (wine steward), Sante Lancerio, cataloged the fifty-seven different wines he and His Holiness had sampled together. While he continued to make regional distinctions, Lancerio was the first to note the color, texture, taste, and aroma of the wines he reviewed, while at the same time introducing words like *round, fat, delicate, powerful, smoky,* and *mature* into the oenological vocabulary.

Modern writing about wine includes all manner of methodologies, ranging from the experiential to the philosophical. In recent years, the focus on environmental determinants such as topography, altitude, climate, humidity, and wind has been especially pronounced in France, where the word *terroir* was coined to encompass the natural geography of viticultural sites. The earliest of these studies tended to concentrate on climate and soil, but lately a handful of articles in specialized journals have dug deeper into the ground to examine the underlying geology of particular regions. Charles Frankel's *Terre de Vignes*, here translated as *Land and Wine*, is the first book-length study written by a professional geologist to take up the subject of how mute stones are indirectly capable of producing human sensations. A celebration of both science and art, the book demystifies the perplexities of wine—and the pretensions of so much wine writing—in a manner as refreshing to the reader as tasting a crisp bottle of Sancerre while learning that its origins lie deep in Jurassic, Cretaceous, and Tertiary bedrock.

John Varriano

PREFACE

Many excellent books about French wines and vineyards have already been published. This book, however, offers a new approach in that it focuses on the influence of terroir—the combination of geography, geology, and climate that gives each wine its special character—and reveals the geological framework behind each great wine region. In Burgundy, we will learn how much the wine there owes its reputation to Jurassic sediment and coral reefs; in Provence, we will explore its vineyards' clay beds, once trampled by dinosaurs; and at Châteauneuf-du-Pape, we will trek across cobble-studded terrain once pounded by the surf of an ancient sea.

Terroir exerts a strong influence on the bouquet, body, and aging potential of any wine. Soil chemistry comes naturally to mind in this regard, from the nourishing elements pumped into the vine by its roots to those involved in bacterial metabolism and the production of aromatic molecules. We will thus explore the rivalry between limestone and clay; note the grain of salt provided by gypsum; and contemplate the mysterious influence of manganese at Moulin-à-Vent. On a larger scale, the size of pebbles and the texture of bedrock play just as important a role, be it the sun-warmed cobbles that restore heat to the vines after sundown, or the water drainage and deep root penetration made possible by rock fractures.

The word *terroir* also encompasses the notion of microclimate, likewise influenced by geology. Topography affects winds and controls precipitation. Erosion determines the shape, grade, and orientation of vine-growing slopes, controlling both sun exposure and water runoff.

The topic has long fascinated me, since I live in France and much enjoy wine. As I am a planetary geologist by training, my field of expertise usually covers volcanic lava flows on the moon and glaciers on Mars. By "landing" in the vineyards of France, I discovered a planet foreign to me, and thirsted to learn (in more ways than one) the secrets behind each landscape and terroir. I was charmed by the beauty of the land and by the gracious hospitality and scientific curiosity of its winemakers.

The thread running through this book is that of time. We will explore French vineyards in geological order, from the oldest to the youngest rock strata. Each chapter will bring to light a different episode of the geological history of France and of the Earth as a whole—not only the creation of mountain ranges and rift basins but the evolution of life as well, since many precious fossils are to be found in the vineyards.

Our time travel shall begin 500 million years ago, in the mica schists of the Anjou region, where the little-known appellations Savennières and Côteaux-du-Layon deliver top-rated wines (chapter 1); continue in Beaujolais with the rise and decay of a great European mountain range (chapter 2); then cover the wearing down of the landscape into flat plains in Alsace, roamed by the Earth's first dinosaurs (chapter 3). The invasion of tropical seas during the Jurassic period leaves a record of oyster beds and coral reefs in the Mâconnais region of Pouilly-Fuissé (chapter 4); in the Côte d'Or of Burgundy (chapter 5); and at Sancerre in the upper Loire valley (chapter 6). This is followed by yet more marine episodes during the Cretaceous period, recorded in the Touraine region of the central Loire valley, famous for its fruity red wines of Chinon, Bourgueil, and Saumur-Champigny (chapter 7). The inland sea then backs out of France, replaced by floodplains and marshes in Provence (chapter 8) and Languedoc (chapter 9), home to the last French dinosaurs. After the great mass extinction that wiped out the giant reptiles, life flourished anew in the terroir of Champagne (chapter 10), and streams rushing down from the newly uplifted Alpine and Pyrenees mountains built up terraces of pebbles and cobbles in the Bordeaux region (chapter 11) and Côtes-du-Rhône (chapter 12), hosting such famous wines as Margaux, Château-Lafite, and Châteauneuf-du-Pape.

Consequently, this book can be read in normal linear fashion, especially since it gives the geological history and evolution of French landscapes in chronological sequence. The chapters also can be read in any order, according to the wine region you wish to explore or the specific wine you just uncorked and want to learn more about (there's an index of geographical and wine names for this purpose).

Since my goal here is to tell a story rather than pretend to write an encyclopedia, I move swiftly down the arrow of time without covering all wine regions, since some are redundant from a geological perspective. Difficult choices had to be made, and great vineyards such as those of Savoy and Jura, Cahors and Corsica had to be left out of my account. My deep apologies to the winemakers of all the areas I didn't mention: their wines are no less superb.

Another rule was to stay away from technical terms as much as possible. I have aimed this book at the general reader, so no particular knowledge of geology or winemaking is necessary. I avoided detailed descriptions of soil types, which is an entire branch of science in its own right, and stuck to a global portrayal of the bedrock that gave rise to the soil—which has a greater story to tell. Finally, I attempted to give the history behind each famous wine, with anecdotes involving kings and queens, philosophers and poets. Beyond terroir and geology, winemaking is indeed and foremost the story of men and women. In the complex equation that describes a great wine, there is soil and landscape, climate and grape variety, but above all the winemaker himself (or herself).

I gathered most the information for this book from the websites of various winemaking regions and individual estates—a list of these sites is provided in the bibliography—as well as from several landmark books, including *The Wines and Winelands of France*, edited by Charles Pomerol (Orléans: BRGM Editions, 2003); *French Wine*, by Robert Joseph (New York: DK Publishing, 2005); and the comprehensive and magnificent *Grand atlas des vignobles de France*, edited by Benoît France (Paris: Solar, 2008), where the reader will find detailed vineyard maps to follow my story and back it up (albeit in French) with exhaustive descriptions of each terroir.

I would like to thank all those who welcomed me onto their estate, showed me their terroir, and had me taste their wine. I mention most of them in the text. My gratitude also extends to the geologists and nature lovers who guided me in regions that were foreign to me, namely Jacques Gispert in Provence and Jean-Pierre Tastet of the University of Bordeaux, who took me around Cassis and the Montagne Sainte-Victoire, Saint-Émilion, Sauternes, and Margaux—all fantastic experiences and memories!

I would also like to express my gratitude to editor Jean-Marc Lévy-Leblond of Editions du Seuil, Paris, for suggesting that I write this book, and to Susan Bielstein, Anthony Burton, and Sandra Hazel of the University of Chicago Press for producing the English edition.

Finally, I propose a toast to all creatures, large and small, that inhabited and built for us—through plankton exoskeletons, oyster shells, and broken-up dinosaur eggs—the wonderful terroirs of France. Their contribution to wine lives on.

CHAPTER ONE

SAVENNIÈRES AND OTHER WINES OF ANJOU

You might expect a book on French wines to begin with Bordeaux, Burgundy, or Châteauneuf-du-Pape. These great wine regions will be covered, with due respect, in other chapters. But because we are dealing with geology and terroir, we will pick up a different thread to weave through the French vineyards. We will follow the chronology of the land, from the oldest to the youngest rock layers underpinning the vineyards.

Exploring the vine-growing regions in this order gives us a sense of the geological history of France, a rich history that encompasses volcanic activity and continental rifting, oceanic flooding of the lowlands, and the building of mountains. Ever more conscious of this rich history, winemakers are now naming some of their cuvées after the geological periods featured in their terroir. You can drink and compare "Triassic" or "Jurassic" wine from Beaumes-de-Venise, or even dive into period subdivisions, tasting "Kimmeridgian" or "Oxfordian" Burgundy.

So let us begin in this spirit with the oldest rock types to boast vineyards in France: Brittany's ancient Hercynian massif, bordered to the south by the Loire valley and cropping up extensively in the wine region of Anjou—2,400 hectares (6,000 acres) of vineyards centered on the fluvial city of Angers. One of the best places to spot the old outcrops and bring to light their influence on winemaking is the small renowned terroir of Savennières.

Savennières is a peaceful village nestled on the right bank of the Loire valley, 15 kilometers (9 miles) downstream from Angers (three hours from Paris by car or train). The region is covered with orchards and vineyards that profit from a warm and gentle climate, and is graced with a luminous atmosphere. Joseph William Turner (1775–1850), the "painter of light" who paved the way for the impressionists, captured its essence in diaphanous watercolors.

The village is situated above the alluvial plains, on the flank of a ridge. It boasts a tenth-century pre-Romanesque church and Renaissance mansions with roofs of fine slate. In the surrounding hills, windmills have long lost their wings and silently stand guard over the vineyards.

Savennières's hilltops are the best place for enjoying a wide-angle view of the vineyards cascading down their slopes, encircling the village, and stopping short of the embankment that drops down to the Loire River's alluvial plain. In the distance, the view stretches south across oxbow lakes, swamps, and fields, then past the sparkling river to its distant left bank barring the horizon, where we can spot more church spires and windmills. The left bank of the Loire has star appellations of its own, including Anjou-Villages, Coteaux-du-Layon, and Coteaux-de-l'Aubance.

You can find every type of wine in the Anjou region: light reds made from Gamay grapes and stronger reds made from Cabernet Franc; fruity rosés; straw-colored dry whites and golden sweet dessert wines, both derived from Chenin Blanc; and even sparkling *crémants*. But it is Savennières that tops the regional hierarchy with its spectacular whites made from Chenin Blanc (fig. 1.1). Two of its terroirs even earn the title of Grand Cru, a rare distinction in the Loire valley: Savennières-Coulée-de-Serrant (fig. 1.2) and Savennières-Roche-aux-Moines. A third terroir in Savennières—Le Clos du Papillon—commands equal respect among wine connoisseurs.

All three terroirs occupy spurs facing the Loire valley, built of schist and volcanic rock. The bedrock is testimony to an ancient ocean basin and to its volcanoes that were crumpled up by plate tectonics against what used to be France's coastline, 500 million years ago.

Baked sediment (schist), altered ocean basalt (spilite), siliceous lava (rhyolite), and veins of pure quartz crop up among the vegetation and supply the vineyard with a generous sprinkle of multicolored pebbles. Facing the southeast and the rising sun, the pebbly slopes are quick at draining rainwater and make for a dry, warm, and mineral-rich terroir. The soil cover is thin, and fractured bedrock is reached a mere foot or two below the surface. As vine roots push downward through the cracks

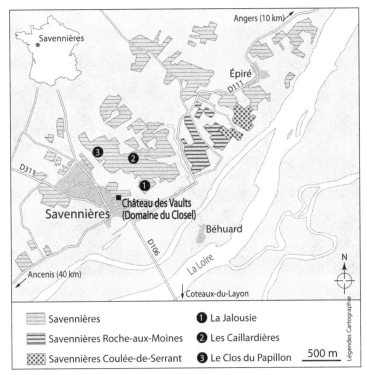

Figure 1.1 The Savennières AOC vineyard, highlighting the two Grands Crus of Roche-aux-Moines and Coulée-de-Serrant, and the three Crus of the Closel estate. (Map by Legendes Cartographie.)

in search of underground water, they ingest metal ions along the way, namely iron and magnesium. Critics often recognize "exceptional mineral flavors" in Savennières wine; one critic, reviewing a particular cuvée of the Domaine du Closel that comes from a lot named La Jalousie, captures its character as "holding the warm and rich notes of a Chenin blanc ripened on schist, with the lingering spices typical of a Savennières."*

To grasp the importance of terroir in bringing greatness and character to wine, we pay a visit to the Domaine du Closel, in the heart of Savennières. Not only does its wine rank among the best in the appellation, it is also astonishingly different from one rock-type terroir to another. Owner Évelyne de Pontbriand encourages visitors to tour her vineyard along footpaths that lead up the slope onto the hilltop overlooking the village. Domaine du Closel owns 15 hectares (37½ acres) of prime-location

*A. Gerbelle and P. Maurange, *Guide des meilleurs vins à petits prix* (Issy-les-Moulineaux: Revue du vin de France, 2005).

Figure 1.2 North of Roche-aux-Moines, the southeastern slope of a rocky volcanic spur is home to the other Grand Cru of Savennières AOC: Coulée-de-Serrant. (Photograph by the author.)

SAVENNIÈRES AOC

Region:	Loire valley, Anjou
Wine type:	dry white
Grape variety:	Chenin Blanc
Area of vineyard:	150 hectares (375 acres)
Production:	4,700 hectoliters/year (627,000 bottles/year)
Famous Crus:	Coulée-de-Serrant
	Roche-aux-Moines
	Clos du Papillon
Nature of soil:	shaly, sandy
Nature of bedrock:	schist, spilite, rhyolite
Age of bedrock:	Silurian (445 million–415 million years ago)
Aging potential:	10 to 20 years
Serving temperature:	10–12°C (50–54°F)
To be served with:	aperitif, fish, shellfish, white meat

vineyards, divided into three separate lots that yield three different nuances of white, based on slight differences in bedrock and microclimate. They are La Jalousie ("Jealousy"), Les Caillardières ("Stony Lot"), and Le Clos du Papillon ("Lot of the Butterfly"), arcing south to west around the hill.

Around the corner from the Savennières church, the Château des Vaults

A SENSE OF HISTORY

Although winemaking in the lower Loire valley dates to Roman times, Chenin Blanc is first mentioned in the Anjou region in the ninth century AD, under the reign of Charlemagne. Monasteries were spreading throughout the Carolingian Empire at the time, and the monks brought with them new grape varieties and savoir-faire. Central to the area was a Cistercian monastery towering over Savennières. The abbey still stands, on a steep volcanic spur of rhyolite, the site and its vineyards aptly named La Roche-aux-Moines ("The Rock of the Monks")—one of the two Grands Crus of Savennières.

There is a historical dimension to the site, since in medieval times monasteries were frequently lined with military fortifications, and La Roche-aux-Moines was no exception. It played a pivotal role in the struggle for land that pitched the nascent kingdom of France, ruled by Philip Augustus (1165–1223), against the kingdom of England, allied with Germany's Holy Roman Empire. In the year 1214, the English army, led by King John, landed on the Atlantic coast and marched toward Paris—a ruse meant to lure Philip Augustus and his modest army from the capital to counter him, and allow the German army to move in from the east. The plan was foiled when the monks and Anjou soldiers barricaded at La Roche-aux-Moines put up a heroic resistance, barring the way to the English invaders; Philip Augustus needed to dispatch only half his men to rout John's army and capture their siege engines. The French king was thus able, with the other half of his army, to counter the true threat marching in from Germany and win a decisive battle at Bouvines, on July 27, 1214. The victory gave Philip Augustus undisputed control of Anjou and Touraine, Brittany, and Normandy.

Once peace returned to La Roche-aux-Moines, the monastery could focus on more spiritual matters and on its wine, which developed quite a reputation within the burgeoning kingdom of France. King Louis XI (1423–1483) called it "a drop of gold." The Sun King, Louis XIV (1638–1715), on a visit to the château de Serrant (the other Grand Cru of Savennières, next to La Roche-aux-Moines), was so impressed by the wine that he wished to visit the vineyard, but the story goes that his coach got stuck in a rut.

Under the French Empire, Napoleon's first wife, Joséphine de Beauharnais (1763–1814), was also very fond of Savennières and made it popular at court. The grandson of Napoleon's chamberlain, the marquis of Las Cases, inherited the château des Vaults at Savennières, along with its vineyard, and his son-in-law Bernard du Closel, mayor of Savennières from 1919 to 1956, was instrumental in developing the reputation of the wine, which was granted formal AOC status in 1952.

serves as headquarters for Domaine du Closel. The elegant sixteenth-century mansion, restored and enlarged around 1830, is surrounded by rosebushes and alleys of cypress and plane trees leading to a little duck pond at the far end of the garden. Should you visit, you are welcomed into a hallway full of antique furniture and then down a few steps to the wine cellar, where Madame de Pontbriand welcomes her guests and shows off her wine.

On a table to the side, three tall vases display the rock and soil types representative of each one of her lots: purple schist at La Jalousie, a touch of siliceous sand over mixed rock at Les Caillardières, and a blend of green schist, altered basalt, and quartz from the highly select Clos du Papillon. A bottle from each terroir stands in front of the chalice of rocks to which it owes its unique character.

The three wines are so distinct that you might even suspect different grapes, but Chenin Blanc is common and exclusive to all. This grape is a variety well adapted to the Anjou climate. It is a late bloomer, waiting out the last frosts of winter before budding and growing leaves, and quickly reaching maturity during the short summer, so that it can be harvested before the early frosts of fall.

Since Chenin Blanc is the sole grape variety on the domaine, this leaves only terroir and microclimate to explain the palette of aromas that differentiate one lot from the next. The steep slope of La Jalousie faces the sun from morning to late afternoon, and experiences torrid temperatures in the summer—as jealousy would have it—whereas the Clos du Papillon benefits from a whiff of cool air on the crest of the hill. But can climate alone justify differences in color, aromas, and flavor between such close neighbors?

The differences are marked indeed. La Jalousie is golden yellow with a glint of green, gives off a bouquet of citrus fruit and lime tree blossoms, and delivers a dry, mineral flavor with hints of white peach, fern, and jasmine. It is served with perch and pike in white butter, crawfish and crab, and fresh vegetables like artichoke and asparagus.

A few hundred meters away, on the flank overlooking the village, Le Clos du Papillon stands out from its neighbor by boasting a deeper golden hue with straw-yellow streaks; a complex bouquet of citrus, lime tree, apricot, and toasted almond, and flavors of lime and hazelnut, jasmine and vanilla, as well as a lingering mineral note of "warm schist." Wine from this lot, Le Clos du Papillon, is rated highly and served in fine restaurants across the country and internationally, where it is recommended with seafood delicacies such as lobster and scallops, turbot and monkfish.

La Caillardière, the third, sandier lot of Domaine du Closel crowning the hill, yields a wine that has the deepest golden hue of all three, giving off a scent of white flowers and aromas of lychee, honey, and acacia blossoms. Wine stewards recommend it with a seafood casserole—such as prawn *au gratin*—as well as with fresh salmon and sushi, or with roasted veal in mushroom sauce.

So many facets for a single wine are difficult to explain based on micro-climate alone, although slope steepness and orientation do play a role. Some aromas are brought in by the winemaking technique itself, such as the vanilla contribution of the oak barrels in which the wine is aged. Decisive, however, is the influence of bedrock. It is close to the soil surface, drains quickly when the rain falls, and allows deep penetration of vine roots, maximizing the area of mineral exchange with the plant.

A LOOK AT SOIL AND ROCK

One way to visualize the complexity of the terroir is to take a walk through the Closel estate, starting at the Château des Vaults and following the path lined with bald cypress and plane trees to the foot of the slope leading up into the vineyards (fig. 1.3). A sign posted for visitors points out that the steep embankment is a good place to view in cross section the topmost layers of the terroir.

The top meter of soil shows distinct layering, starting with a humus horizon about 10 centimeters (4 inches) thick (the "A horizon" known to specialists), made up of leaf litter and other organic matter that is being actively processed by bacteria and insects. Beneath this top layer lies a transition zone of roughly equal thickness, where organic matter gets mixed with mineral particles from below—a gardening task provided free of charge by burrowing worms and insects.

About a foot below the surface, the soil is mostly mineral but lacks its most soluble elements, such as iron and aluminum, which rainwater has leached out and transported downward. Fortunately, much of this scavenged metal ends up in a lower "B horizon" that is within easy reach of the vine's root system. This layer concentrates iron, aluminum oxides, clays, and other minerals. Deeper yet is the transition zone with the bedrock—broken-up rubble known as the "C horizon" or regolith—and then the bedrock itself, a couple of feet below the surface at this location.

You might think that bedrock raises an impenetrable barrier to vine roots. If this were so, however, the vineyard would do very poorly indeed, because two feet of permeable topsoil do not retain much water during the hot and dry summer months. It is thanks to the fractured bedrock that the grapevine can reach water and nutrients at depth.

At Savennières this bedrock is schist: a pile of sediments cooked under pressure in the Earth's crust long ago and brought back to the

Figure 1.3 A trail for visitors to the Closel estate winds through the vineyard and its three "climates," each of which possesses distinct soil and bedrock. (Map courtesy of Closel estate.)

surface today.[†] Once it cools, schist becomes brittle and breaks up into sheets. Its many fractures allow the vine roots to penetrate the rock to

---

[†] Schist is classified as a *metamorphic* rock, since it underwent physical and chemical changes (metamorphism) during its burial in the Earth's crust, under conditions of high temperature and pressure. Slate is another form of metamorphic rock, slightly less cooked and altered than schist.

Figure 1.4 *In the foreground*, the vineyard of Grand Cru Roche-aux-Moines; *past the tree line*, the Jalousie "climate" of the Closel estate; and *on the horizon*, the Caillardières "climate" and the white buildings of Moulin-du-Gué. (This and all other photographs not otherwise credited are by the author.)

depths of 7 or 8 meters (23 or 27 feet), where it can tap water. As a bonus, the long path followed by the roots makes for a greater surface contact with the bedrock and hence a proportionately greater mineral exchange with the vine.

We can spot the slaty bedrock in the embankment, and we can trace it uphill under the thin soil cover as we walk up the path out of the underbrush and into the sun-drenched vineyard (fig. 1.4). As the path rises above the village, we then catch a view of rooftops covered with fine bluish slate—a form of schist used in roofing and decoration throughout Anjou.

At the foot of the vines, chunks of schist mix with quartz pebbles and many other colorful stones—pink and white, green, purple, and black—that give the Savennières terroir the appearance of an open-air candy store. These mineral nuggets soak up sunshine during the day and radiate their stored heat at night, serving as miniature radiators that protect the vines from nighttime cold spells. And of course, these same pebbles—along with their parent bedrock—give Savennières wine its celebrated mineral flavor.

THE MAKING OF FRANCE

Besides the role terroir plays in nourishing and heating the vineyard, rock fragments here have a fascinating story to tell: the very birth of France as a microcontinent, 500 million years ago. By delving 6 or 7 meters (20 or 23 feet) underground, through purple schist, green schist, and veins of quartz, the roots of Chenin Blanc are tapping a tumultuous history of closing ocean basins and colliding continents.

Half a billion years ago, the Earth had already been around for a long time — 4 billion years, in fact — but its landscapes were still barrenly mineral, without any sign of life. Life was restricted to the oceans, where it was undergoing a surge from simple colonies of algae and ancestral jellyfish to much more complex marine animals — a surge known as the Cambrian explosion, named for the geological period when it took place. By 500 million years ago, sponges blossomed and elaborate arthropods known as trilobites crawled and fed on the sea bottom, and the first free-swimming, fish-like vertebrates were fast developing.

During the Cambrian, landmasses were placed differently than they are now, since plate tectonics have constantly been at work opening up new ocean basins, closing old ones, and jostling continents around. A giant ocean covered the Northern Hemisphere at the time, and all the continents were packed together in the Southern Hemisphere, most of them locked into an unrecognizable jigsaw puzzle. Africa and South America were spliced together and turned upside down, and South Africa and Patagonia were pointing north. Off to the side, North America, Siberia, and the Baltic lands were separate continents, each drifting its own way.

As for France, it was yet to be assembled. It existed as three different plate sections. The piece that would end up becoming northern France (the Ardennes, Picardie, and Nord-Pas-de-Calais regions) was welded to the Netherlands, northern Germany, and England, forming a continent named Avalonia, in reference to the mythical island of King Arthur and the Knights of the Round Table. Another small plate to the west comprised Brittany and part of Normandy: geologists refer to it as the Armorica block. Third, the southern half of France was attached to the great continental block of Africa and South America.

These tectonic plates were moving relative to one another at rates that rarely topped 10 centimeters (4 inches) per year. In places, underwater rifting ripped up the ocean floor, adding slices of fresh magma to the expanding crust and driving continents farther away from one another. Elsewhere, ancient colder and denser crust bent downward and sunk back into the hot mantle, shrinking ocean basins and bringing distant shorelines head to head.

The overall picture 500 million years ago was one of change. A small ocean had been steadily growing between the Armorica microcontinent—Brittany and Normandy—and the block of southern France attached to Africa. Submarine volcanoes built up on the expanding seafloor, and deepwater sediments piled up on top of the lava—an ephemeral basin named the Central Ocean by French geologists, since the site is now occupied by the Massif Central (fig. 1.5). Life was abundant in this rapidly evolving ocean basin. Bottom-dwelling organisms built up reefs crawling with trilobites—segmented arthropods that resembled the wood louse and are prized by fossil hunters—while in the open waters giant cepha-

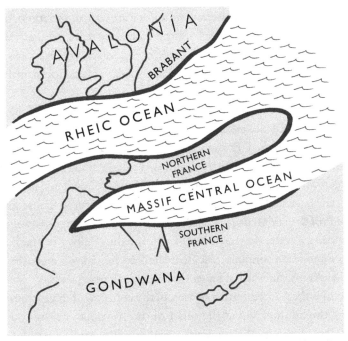

Figure 1.5 France during the Ordovician period, 450 million years ago. The future terroir of Savennières belongs to the ancient Massif Central ocean, which was closed up by the convergence of tectonic plates along a fault zone that today is followed by the Loire River. (Map by Pierre-Emmanuel Paulis.)

lopods, distant ancestors of squid and octopi, cruised in search of prey, encased in slender shells that were the shape and size of torpedoes.

Primitive fish also haunted the scene: awkward-looking, clumsy proto-types with bony plates for protection, unsophisticated rudders for direction, two bulging eyes sticking out like flashlights, and a mouth shaped like a vacuum cleaner, lamprey-style.

This Central Ocean and its menagerie took a turn for the worse around 400 million years ago (a geological period known as the Silurian), when plate tectonics reversed direction and started driving most continents northward, albeit at different rates; the southernmost fragments caught up with the frontrunners and swallowed up the ocean basins in between. Jammed between northern and southern France, the Central Ocean basin bent downward under the converging continents and inched its way back into the hot mantle, its area progressively shrinking. This road to hell was paved with bouts of melting and volcanic eruptions, and the sediments scraped off the condemned ocean plate piled up along the continental margins. As these eventually converged and closed in jaw-like fashion, swallowing the last bits of ocean crust, the northern and southern halves of France were welded together along a fault zone that later materialized as the lower Loire valley. The tectonic collision spared slivers of ocean basin, thrusting and mingling sedimentary and volcanic rocks into what would eventually become the Savennières terroir.

When you uncork a bottle of Savennières, it is the spirit of this vanished ocean that you are releasing: the memory of a prehistoric collision that gave birth to the French microcontinent.

It is no coincidence that the Loire River flows along this suture zone between northern and southern France. Rivers follow fault lines when they can, since faults juxtapose rocks of different natures. Consequently, one side of the fault is usually harder than the other, acting as an aqueduct to funnel the water down the line. The Loire thus follows the fault zone, as do some of its confluents, such as the Layon, which happens to host another Anjou vineyard of fame: Coteaux-du-Layon (see box).

Besides the faulted terroir of Anjou, the collision of microcontinents during the Silurian and at the onset of the Devonian periods built up most of the geological framework of France. It was also a time of revolution in the plant and animal kingdoms, since life left the oceans and tentatively set foot on dry land. Until this crucial turning point, living

COTEAUX-DU-LAYON: A VOLCANIC TERROIR

The assembly of microcontinents to form France 400 million years ago involved many different types of volcanic activity. There were peaceful lava flows, made up of fluid basalt, which have since been altered to green and bluish rocks named spilite that outcrop throughout the Savennières terroir. There were also explosive eruptions involving viscous, colorful lava known as rhyolite. These disturbances built up mounds that poke up in today's landscape, be it at Savennières proper, where they make up the famed terroir of La Roche-aux-Moines, or else on the left bank of the Loire in the rocky hills that host the Coteaux-du-Layon vineyards—another renowned appellation of Anjou.

A confluent of the Loire, the Layon River follows a northwest-striking fault and joins the main river at Chalonnes-sur-Loire. Along the way, it is fed by its own set of smaller confluents that cut up its left bank into a festoon of vine-planted hills. Here the hot summer sun brings full ripeness to Chenin Blanc—the same grape variety as at Savennières—but morning fog clings to the hill slopes in the early fall, promoting the growth of a microscopic fungus on the grapes, *Botrytis cinerea*, also known as "noble rot." The fungus drains the water from the grapes and concentrates their sugar, yielding a sweet dessert wine akin to Sauternes, but less pricey. Coteaux-du-Layon wines are served as an aperitif, with foie gras, or at the end of a meal with chocolate desserts.

Some Coteaux-du-Layon wines have a flinty, smoky aroma that is credited to the volcanic rhyolite outcropping in many lots, especially between the villages of Beaulieu-sur-Layon and La Faye: you can spot the siliceous, taffy-like pink lava in the embankment of road D 55 along the vineyards. Closer to the Loire, at the entrance of the village of Rochefort, you can admire a nice outcropping of rhyolite in a small abandoned quarry at the start of a little road leading up to Pic Martin—the hill overlooking the village. The quarry face is the best place to see the rhyolite exposed, stripped of soil and vegetation, displaying its folded flow textures and orange, pink, and red hues.

Elsewhere in the Coteaux-du-Layon district, you can spot fluid basaltic flows, probably remnants of the Central Ocean floor (or of a "back-arc" basin that developed behind it, as is common in present-day subduction zones along the Pacific rim). An impressive open pit slices through the lava layers off route D 55 near the town of Beaulieu. The active quarry front is close to 100 meters (330 feet) deep and exposes dark basalt, altered into green spilite rock by millions of years of hydrothermal brewing. Geologists allowed in the pit can view the rock pile up close and trace the bulbous shapes of "pillow lavas"—a telltale sign that the magma erupted underwater on the seafloor.

With its black basalt and red rhyolite, the Coteaux-du-Layon terroir has a rich and varied volcanic past, marking the extension and then the closing and compression of the oceanic frontier between northern and southern France, 400 million years ago.

organisms had remained underwater—the environment in which they were born—colonizing the oceans and slowly moving up brackish estuaries and freshwater rivers. Continents were totally bare: desert landscapes of rock and sand, akin to the red planet Mars. Only in the Silurian and Early Devonian, 400 million years ago, did moss and lichen creep up the riverbanks, followed by early insects, vascular plants with stems and leaves, and finally the first amphibians that branched off from the fish family and crawled out of the water on fin-like limbs.

We are still hundreds of millions of years away from sophisticated rampant vines of Chenin Blanc, hares dashing through the rows and birds chirping in the bushes of Savennières, but evolution was now on a roll. We will follow its many twists and turns as we proceed to explore the terroirs of France.

CHAPTER TWO
BEAUJOLAIS

On the third Thursday of November, cafés and wineshops around the world uncork the brand-new production of Beaujolais Nouveau: a celebration of the not-so-distant harvest—held two months prior—and a brilliant marketing ploy, not only to publicize the Beaujolais brand but also to rake in early cash for the winemakers.

Celebrating the harvest and tasting the new vintage is a tradition that goes back to the dawn of civilization, but regulations in France had long set the release date of new wines for December 15. When an exception to this rule was decreed in 1951, moving up the date by nearly a month, the partying that ensued in bistros spread from Lyon to Paris and eventually to the entire world, when the dean of Beaujolais winemakers, Georges Dubœuf, coined the successful marketing slogan "Le Beaujolais Nouveau est arrivé" (Beaujolais Nouveau has arrived) and aggressively flooded the market with young wine. Today, nearly 40 million bottles of Beaujolais Nouveau are shipped every November to over one hundred countries, the largest consumers after France being Japan (6 million bottles) and the United States (2.3 million bottles).

The two-month-old Beaujolais is a pleasant, light, and festive wine. Its aroma is often artificially spiked with "designer yeasts," bringing out flavors such as banana (yeast 71B), raspberry, or bonbon candy. Under these circumstances, you wouldn't expect the terroir to have any say in its personality. But compare from the same producer, if you can, a Beaujolais Nouveau and a Beaujolais-Villages Nouveau. You will find the latter to have more of a mineral bite, which it holds from the rockier, higher-quality terrain on which it is grown. But if you truly want to explore the subtleties of terroir, drink up your juvenile Beaujolais Nouveau and move on to the true, mature wines of the region.

Beaujolais has indeed much more to offer than its Nouveau wine. The region offers a whole gamut of reds, ranging from light and fruity generic Beaujolais to more complex Beaujolais-Villages, ten of which have earned special Cru status and are known by their village name alone—like Moulin-à-Vent, Morgon, and Saint-Amour—to the point where most Frenchmen have forgotten that these too belong to the Beaujolais family.

Geographically speaking, stretching across the departments of Rhône and Saône-et-Loire, the Beaujolais region is the southern extension of the region of Burgundy, from the town of Mâcon to the suburbs of Lyon, over a distance of roughly 50 kilometers (30 miles). Along this north–south axis, the vineyard is a dozen kilometers (7 to 8 miles) wide on average, climbing from the alluvial plain of the Saône River up into the western piedmont hills of the Beaujolais mountain range.

Besides a slightly warmer climate, the big difference between Beaujolais and Burgundy is the nature of their terroirs, favoring different grape varieties. In Burgundy limestone prevails—a rock type made up of tiny shell fragments and other carbonate material. Pinot Noir and Chardonnay do very well in this type of soil, producing great wines. But south of Mâcon, there is an abrupt change in bedrock. The Beaujolais hills switch to a granitic and volcanic terrain that favors another grape variety, known as Gamay.

The granite and lava rock break down into mineral-rich gravel and sand, nourishing a range of "crus" that each display a distinct personality, inherited from the intricacies of the terroir and enhanced by the local winemakers. From north to south, the ten villages that have earned Cru status are Saint-Amour, Juliénas, Chénas, Moulin-à-Vent, Fleurie, Chiroubles, Morgon, Régnié, Brouilly, and Côte-de-Brouilly (fig. 2.1).

Their reputation goes back a long way. Some terroirs were already famous in Roman times, as was the hill of Brouilly, named for Brulius—an officer of Julius Caesar's army who received vine land from the emperor in exchange for his loyal service. As for the town of Juliénas, it was purportedly named for Caesar himself.

The vines planted in Roman times, first introduced in the area by Celtic tribes, were probably Allobrogica—the presumed ancestor of Pinot Noir.

Figure 2.1 The ten renowned Crus of Beaujolais, on hills overlooking the Saône River. (Map by Legendes Cartographie.)

Later barbarian invasions allegedly brought in the Gouais variety, which, crossed with Pinot, delivered two successful hybrids: Gamay and Chardonnay. Today, Pinot Noir, Gamay, and Chardonnay are the three dominant grapes of Burgundy and Beaujolais wines.

The Gamay variety does very well in temperate climes. It ripens earlier than most other varieties and is extremely prolific, which appears to be an advantage. But the abundant grapes do not yield good wine everywhere, and in the Middle Ages Gamay was judged severely by the ruling nobility. Well aware of the delicate balance needed to match grape and terroir, Philip the Bold (1342–1404), Duke of Burgundy, recognized the superiority of Pinot Noir over Gamay on his land and ordered his winemaking subjects, in a decree dated July 31, 1395, to "uproot, destroy and set to waste," as far south as Mâcon, "the bad and disloyal plant of gamay, a plant that yields great quantities of wine . . . , wine of such nature that it brings much harm to human creatures, for it is full of tremendous and horrible bitterness."

The review is certainly a bit harsh, but we get the point: cultivated on Burgundian limestone, Gamay is disappointing. The 1395 eradication law, however, applied only to the vineyard north of Mâcon. Winemakers continued to grow Gamay south of the city, where it fared much better on granitic soil up in the Beaujolais hills. The grape was also introduced in the Loire valley to the west and in a few other regions, including parts of Switzerland; but to this date, two-thirds of the worldwide production of Gamay wine still comes from the Beaujolais area.

BEAUJOLAIS AOC

Region:	Beaujolais (Southern Burgundy)
Wine type:	red
Grape variety:	Gamay
Area of vineyard:	10,000 hectares (25,000 acres) (Beaujolais)
	6,400 hectares (16,000 acres) (Beaujolais-Village)
	6,200 hectares (15,500 acres) (Crus)
Crus:	Brouilly, Chénas, Chiroubles, Côte-de-Brouilly, Fleurie, Juliénas,
	Morgon, Moulin-à-Vent, Régnié, Saint-Amour
Nature of soil:	shaly, sandy (uphill and in the north)
	clayey and calcareous (in the south)
Nature of bedrock:	schist, granite, amphibolite (uphill and in the north)
	limestone (in the south)
Age of bedrock:	Carboniferous (350 million–300 million years ago) in the north
	Mid-Jurassic (Aalenian, 175 million–170 million years ago) in the south
Aging potential:	1 to 3 years (2 to 8 years for Crus)
Serving temperature:	12–15°C (54–59°F)
To be served with:	cold cuts, chicken, veal

PINK GRANITE AND GOLDEN LIMESTONE

Different grades of Beaujolais correspond to different terroirs. On the one hand, there is the flat land we see from the highway (A6) between Mâcon and Lyon: the valley of the Saône River, with the hills a distant backdrop. Here the terroir is made up of recent sediment, mostly alluvium brought

FROM CLUNY TO BEAUJOLAIS NOUVEAU

Vine planting in Beaujolais goes back to Roman times, but the region truly developed in the Middle Ages, under the influence of the lords of Beaujeu—a fortified city in the Ardières valley that gave its name to the region. The alliance struck in the tenth century between the lords and the Benedictine monks of the Cluny order warded off the ambitions of their powerful neighbors—counts and bishops of Mâcon and Lyon.

The winemaking know-how of the Cluny monks, and later of the friars of the Abbey of Belleville founded by the Beaujeu lords, helped expand and improve the vineyard. The establishment of a free trading town on the Saône River in 1140—Villefranche-sur-Saône—improved on the other hand the transportation and trade of wine, particularly southward toward the city of Lyon. According to an old saying, three rivers converge in Lyon: the Rhône, the Saône, and Beaujolais wine . . .

Light, fruity Beaujolais is a natural match for Lyonnais cuisine, cold cuts, and cheeses. In mid-nineteenth century, the creation of the Paris-Lyon railroad, which passes through the Beaujolais region, opened up a new trade route to Paris. It is no coincidence that Georges Dubœuf, the main producer of and marketing whiz behind Beaujolais wine, installed his wine museum, Le Hameau du Vin, in the old railroad station of Romanèche-Thorins, gateway to the Grands Crus Fleurie and Moulin-à-Vent.

It is also Dubœuf, along with writer and agronomist Louis Orizet, who in the 1970s launched the promotion of Beaujolais Nouveau, a two-month-old wine released on the third Thursday of November. From 2 million bottles in 1951, the year when wine-controlling authorities first allowed such advance sales, the production of Beaujolais Nouveau has reached 100 million bottles today, roughly a third of the total production of Beaujolais wines.

in by repetitive flooding of the valley, and partially loess—windblown sediment. This silty clay loam yields the lightest Gamay wines, which are granted the regional appellation Beaujolais, without any suffix or village name. Pleasant and easy to drink, they are known for their mildness and fruity aromas of ripe grape, cherry, or raspberry.

The southern tip of the Beaujolais region, south of Villefranche-sur-Saône and down to the outskirts of Lyon, has limestone strata poking up through the alluvium, witness to an ancient Jurassic sea. Gamay wine from this calcareous terroir is equally light and fruity, pleasant and thirst quenching, but with a little more bite and harshness to it, credited to the limestone.

These marine sediments date to the Jurassic period, 200 million to 150 million years ago, and yield a fine building stone with a golden hue, known in French architecture as *pierre dorée*. Most monuments and traditional houses in the region display the fine limestone—as does the village of Oingt in the Azergues valley, listed as one of the most charming villages in France, with ramparts, church, and castle all built of the golden stone. The *pierre dorée* is also featured in Beaujeu, capital of the Beaujolais region, and in many of the winemaking villages.

As we leave the lowlands to climb up into the foothills, we cross over buried faults to enter a rocky province 100 million years older, crystalline and magmatic in origin. This is the granitic and volcanic terroir that hosts the Beaujolais-Villages appellation and its ten top-of-the-line Crus, and gives Gamay an entirely different dimension. In contrast to Beaujolais wine raised on clay and limestone, Beaujolais-Villages and its Crus are denser, more structured, with stronger personalities and greater aging potential.

The granite hills of Beaujolais pick up the geological history of France where we left off in Anjou, with mighty plate collisions building up the French microcontinent and giving rise to volcanic lavas and metamorphic schists. We saw how one collision closed up a Central Ocean along a suture zone, materialized today as the east–west fault system of the lower Loire valley. A few million years later, another plate collision affected the area to the north, crushing a small ocean basin to yield a second suture zone, outlined by the flexure zone of the English Channel.

As the French territory got spliced together in this manner around 350 million years ago, mountains rose along the clashing borders, as tall and spectacular as the Himalayas are today. The pan-European range, named the Hercynian massif, is actually an extension of the Appalachian Mountains of North America (fig. 2.2). In Europe, these tall mountains have long since vanished, destroyed by erosion, but their worn-down roots can still be seen as folded gneiss and schist in Brittany, the Vosges and Ardennes hills of northeastern France, the Rhine Massif of Germany, and the Bohemian Massif of Poland and the Czech Republic.

This glorious period of high mountains coincided with the Carboniferous period: dense forests covered the lower slopes, and lakes and swamps filled the rift basins at the foot of the range, teeming with insects, amphibians, and early reptiles. Growing like two decks of cards merging together, the Hercynian chain eventually became unstable and spread outward under its own weight, vigorously eroding and creeping along strike-slip faults.

As the chain collapsed, the deep mountain roots were stripped of their

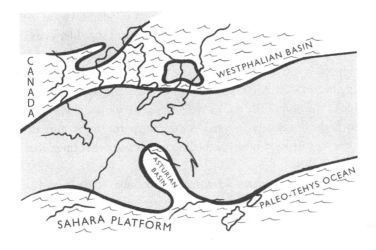

Figure 2.2 France during the Carboniferous period, 310 million years ago. The entire country is above water, including the site of the future Beaujolais vineyard, and belongs to a wide Hercynian continent crossed by a mountain range. (Map by Pierre-Emmanuel Paulis.)

massive load, experienced declining pressures, and began to rise like hot-air balloons, melting on their way up. Pods of granite stalled close to the surface and volcanic eruptions broke out into the open—the swan's song of the Hercynian massif. The last remnants of its bygone splendor are the subdued rolling hills that today carry the ten celebrated Crus of Beaujolais.*

BROUILLY, GATEWAY TO THE BEAUJOLAIS HILLS

The present-day Beaujolais landscape owes its rolling hills not so much to the long-leveled ancient mountain range as it does to a much more recent uplift: an aborted rift zone that spread north to south through eastern France a mere 35 million years ago, bulging the crust along old faults and dropping down the central axis as a series of rift valleys: the Alsatian rift (Fossé d'Alsace) in the north; the Saône valley and Bresse region in the middle; and the Limagne branch in the south.

* If you visit the area, you can learn more about the geological history of the Beaujolais region at the Espace Pierres Folles Museum, which also hosts a botanical garden and a nature trail. Address: 116 chemin du Pinay, 69380 Saint-Jean-des-Vignes. Phone: 04 78 43 69 20. Website: www.espace-pierres-folles.com/.

As the rift basins subsided, their margins rebounded skyward to create rift shoulders. In the Bresse region, the western rift shoulder even claims mountain-range status, with the Monts du Beaujolais reaching 1,000 meters (approx. 3,300 feet) in height. The high hills intercept the damp air blowing in from the Atlantic, with most of the rain falling on the western side of the range. The cooler, dehydrated air blows down the eastern flank over the Beaujolais vineyards—a dry foehn that fights humidity and keeps the vine in good health, repelling molds and other fungi that need humidity to thrive.

Reinforcing this climatic advantage, most slopes face east and heat up early in the day. Those facing south also get great sun exposure. As for slopes facing southwest to due west, they lack sun in the morning and make up for it in the afternoon, but the number of sun hours can be considerably shortened in the evening by long shadows cast over the vineyard by the higher mountains of the range.

Improving the terroir even further, the front of the Beaujolais range is sculpted by erosion into a wavy front of piedmont hills—the work of many rivulets flowing down to the Saône River. As a result, the hill front is carved on all sides, offering every possible orientation with respect to the sun.

One of the best examples of such a detached hill is Mont Brouilly. It is often mistaken for a volcano because of its rounded shape and steep slopes. Strictly speaking, it is not a volcano, though made up of ancient volcanic rock. Its peculiar shape is simply the result of erosion.

The Brouilly hill and the plains around it are planted with vine (fig. 2.3). The steeper slopes, above 200 meters (approx. 650 feet) in elevation, yield a Cru called Côte-de-Brouilly. The lower slopes deliver a wine simply called Brouilly, by far the vastest of the ten Crus of Beaujolais, with 1,300 hectares (3,250 acres) planted. The garnet-red Brouilly is a rich fruity wine that develops aromas of cherry, red currant, or blueberry, a floral touch of peony, and hints of coffee and even licorice.

Expert wine tasters will pick up slightly different characters of Brouilly wine, due mainly to location around the hill and the terroir that goes with it. On the western side, the terroir is granitic, covered with pink pebbles rich in silica. The eastern side is on the contrary calcareous—a limestone terroir that yields lighter wine. The southern side consists of blue volcanic rock shed from the higher slopes of Côte-de-Brouilly and shares some of the characteristics of the latter.

Figure 2.3 Harvesttime in the Côte-de-Brouilly vineyard, on the southern flank of Mont Brouilly.

CÔTE-DE-BROUILLY: A VOLCANIC CRU

The high reaches of Mont Brouilly deserve their classification as a separate appellation. The slopes reach 40 degrees in places, and the summit is covered with trees. Below the tree line, the hill is planted on all sides with vineyards. Even the northern slope, sculpted over the millennia by the Ardières River, is steep enough to make up for its poor exposure and receive the sun's rays at a decent angle. Altogether, 310 hectares (775 acres) girdling the mountain are classified as Côte-de-Brouilly.

On the rounded summit, a little chapel was built in 1857 to implore protection against vine-threatening diseases and fungi such as powdery mildew (*oïdium*). Not so long ago, processions of devout villagers marched up the steep hill to the chapel, beseeching mercy from heaven.

Côte-de-Brouilly wines have a deeper garnet-red color than their Brouilly neighbor downhill. They have a powerful bouquet of red berries, black cherry, and spices, with sometimes a whiff of iris. The flavor lingers nicely, and Côte-de-Brouilly ages well—a challenge to the preconceived idea that all Beaujolais should be drunk young. Here, you can keep a well-produced wine for three or four years, or even longer if the vintage is right.

Which brings us to the influence of terroir: Côte-de-Brouilly benefits from a special kind of rock that plays a leading role in its character—and makes the hill on which it is grown pass for a volcano. This bedrock is altered diorite, a bluish magmatic rock outcropping in the vineyards and in the woods above them, showering the slopes with blocks and smaller fragments.

The *pierre de Brouilly*, as it's called, is very handsome. Blue with green-ish streaks, it is hard and dense, and was mined in the late 1800s to provide ballast for roads and railroads. Pickaxes and rock crushers took a beating, however, and compounded with the steep slopes that made access diffi-cult, the few quarries on Mont Brouilly eventually closed shop.

The blue rock of Brouilly is still used locally in construction. The few houses on the hill—all belonging to winemakers—show it off in their walls, cornerstones, and windowsills, or as raw blocks in their courtyards and gardens. Out in the vineyard, vine workers pick up the rocks and pile them up in rows across the hillslope to slow down soil erosion, walls which go by the local name of *troussées*.

This rock is also very useful in understanding the geological history of the site. Diorite is a rock that crystallizes out of a fairly siliceous magma (around 50 percent silica)—the type of magma that ascends along sub-duction zones where a water-soaked tectonic plate dives below another and partially melts during its descent into the hot mantle. Magma released by this water-rich melting becomes diorite if it slowly congeals below the surface and develops large crystals, or andesite if it erupts at the surface and cools rapidly as glassy lava with smaller crystals.

At the close of the Devonian period, 380 million years ago, gas-rich, explosive volcanoes erupted in thick lava flows and rolling clouds of ash. The diorite of Mont Brouilly is not ash or lava per se but magma con-gealed below the surface in the plumbing system of a volcano that has long since eroded away. The rock is also unusual in that it underwent a second phase of cooking by contact with a hot plume of granite that rose into the area millions of years later. The extra round of baking altered its minerals toward deeper blue and sometimes green varieties, so that the rock is no longer a pristine diorite but more precisely an amphibolite (blue-green amphibole is the major mineral of altered diorite). In addi-tion, milky quartz veins rich in silica run through the rock fractures, giv-ing the rock an elegant marble-like texture.

Volcanologists thus interpret the Mont Brouilly diorite as a legacy of oceanic volcanism, along a tectonic trench full of sediments. Altered by the circulation of hot seawater in its cracks, baked by its burial at depth and by the upwelling of hot granite, the blue rock of Brouilly is notably low in potassium and rich in sodium. The sodic nature of the soil is believed to play a role in both the metabolism of the grape and the wine's special character.

Besides a chemical role, the blue diorite also plays a physical one, heating up the terroir. Because the rocks and stones are dark colored, they absorb sunlight and later give off the heat to the soil and vines, like miniature radiators.

When you drink a Côte-de-Brouilly, then, have a thought for the blue rock. Besides, it is often featured on the wine label as a blue landscape or as blue lettering (fig. 2.4). Several estates even make it part of their name, as does Les Roches Bleues ("The Blue Rocks"), just off the road that

Figure 2.4 Côte-de-Brouilly winemakers often picture the hill and mention its blue rocks on their label, as does Les Roches Bleues. (Photograph courtesy of the estate.)

crosses the southern slope of the hill. Boulders of diorite greet the visitor along the alley leading up to the house, where Christiane and Dominique Lacondemine will have you taste a garnet-colored Côte-de-Brouilly giving off a bouquet of jammy red and black berries and offering the tangy bonbon freshness of Gamay, with good tannin structure and a long finish. Like bona-fide Burgundies, their Côte-de-Brouilly is aged in oak barrels for several months before it is bottled in the spring—or up to a year for the better vintages. These bottles can then be stored several years in a wine cellar, where they will continue to improve over time.

RÉGNIÉ

Régnié is the tenth and last Beaujolais village to achieve Cru status, in 1988. Its 395 hectares (987 acres) of vine are located in the township of Régnié-Durette and are framed by Brouilly to the south and Morgon to the north. The slope climbs steeply westward into the Beaujolais foothills, with the vineyard ranging from 250 to 500 meters (830 to 1,660 feet) above sea level.

At Régnié the underlying rock is pink granite, yielding a terroir of granitic sand and gravel. The permeable, well-aerated soil drains away excess water and favors gas exchange between the atmosphere and the vine roots. Gamay wines from Régnié develop a bouquet of berries—raspberry, red currant, and black currant—and occasionally peach. Tannins are mellow, giving Régnié a feminine character that puts it on a par with Chiroubles and Fleurie. Nonetheless, Régnié from the steepest sun-facing slopes can be full bodied and require several years to develop its full aromatic palette.

MORGON AND ITS ROTTEN ROCK

Mont Brouilly and its volcanic plumbing system give us a picture of what Beaujolais was like 380 million years ago, when volcanoes were involved in a plate tectonic collision that led to a great phase of mountain building in Europe. North across the Ardières River, the neighboring terroir of Morgon picks up the story. It takes us to the closing scenes of the mountain-building cycle, when the great Hercynian range started to collapse and spread outward under its own weight. Crustal spreading and extension opened up new faults and fissure zones, allowing basaltic magma to well up in the basins. This is why we find old, greenish volcanic rock in the Morgon terroir, known as *pierres volantes* ("flying stones"), especially on Côte du Py—a hill in the southern section of Morgon that produces some of the best wine.

Between the veins of altered lava, the most common rock is ancient marine sediment belonging to the same period, which was baked by the magmatic activity and now outcrops as flaky, dark-colored schist, sprinkled with some lighter-toned aluminous minerals. Classified as a pyritic schist, the oxygen-poor rock is good at preserving fossils. Occasionally, it yields leaf imprints and even entire trunks of Carboniferous trees— *lepidodendron* fern trees reaching 20 meters (66 feet) and equally tall, scaly-barked *calamites* horsetail trees.

Fossils from the animal kingdom are much rarer, and none are found in the Morgon area. But from what we know of the Carboniferous period, 350 million years ago, early species of amphibians, looking a lot like present-day salamanders, had crawled out of the waters and started exploring the continent, paving the way for all vertebrates to follow, including future reptiles and mammals.

At Morgon, winemakers have a name for the Carboniferous marine schist that breaks up into small platelets and crumbles to clay. They call it *roche pourrie* ("rotten rock"), but there's nothing rotten about it when it comes to the wine it produces. On the contrary, the rock's many fractures enable the vine to sink its roots several meters underground and profit from the readily available minerals. The Gamay grape has access to interesting elements and compounds such as sodium, and iron and manganese oxides that color the soil shades of red and black. According to winemakers, these oxides are important in the aging process of Morgon.

On the schist-rich Côte du Py, Morgon wine is even qualified as giving off "a bouquet of *roche pourrie*," and oenologists pick out whiffs of gunpowder and black pepper, which they credit to the schist and its metal-rich clays.

When drunk young, Morgon is a fruity wine, garnet colored with a bluish tint, that gives off aromas of pitted fruit such as apricot, peach, prune, and cherry. Morgon from Côte du Py and other schist-rich terroirs are worth aging up to six or seven years: they become fleshy, and their rich bouquet brings to mind cherry pits and cooked fruit. This flavor of ripe cherry or even sherry brandy is typical of the most evolved Morgon wines. Winemakers and oenologists even made up a word to describe this transformation: for them, a Morgon *morgonne* (it's as simple as that!). This schist-raised, oak-barrel-matured Morgon can even exhale a bouquet of ginger and cinnamon, coffee, licorice, or even cocoa.

Winemaker Daniel Rampon, who owns lots on both the celebrated Côte du Py and the more alluvial ground at Les Micouds, notes that there are many subtleties and grades of Morgon wine. Six different terroirs are recognized around the central village of Villié-Morgon. The southernmost terroir, Les Grands Cras, features granitic soil, as do the eastern terroir of Douby and the northern halves of Les Charmes and Corcelette. Closer to the center of the appellation, the celebrated Côte du Py is famous for its "rotten" schist. As for Les Micouds, stretching west in the direction of the Saône River, it features a large amount of alluvial, iron-rich clay.

CHIROUBLES

The Beaujolais Cru of Chiroubles can be reached by following the twisting D 86 road from Villié-Morgon upward toward the northeast. The upper terraces of the hillside offer a spectacular view of the vineyard, the Saône valley, and the Bresse basin in the background.

The Chiroubles vineyard is nestled in a south-facing amphitheater at elevations between 250 and 450 meters (830 and 1,500 feet) above sea level, which allows for a slightly cooler climate than with most other Beaujolais Crus: harvest usually occurs a week later, in order for the grapes to reach maturity. Pink granite shows at the surface, and some of the lots were actually pried open with iron crowbars, although in other places the rock has naturally broken down into a coarse sand locally named *gore*. Rock fragments are piled up to buttress the terraces, slow down erosion, and indicate lot divisions—low walls known as *chirats* that are probably responsible for the town name of Chiroubles.

The granitic terroir, high elevation, and good sun exposure add up to make Chiroubles one of the most fragrant and enjoyable Beaujolais Cru. It is described as feminine, meaning fragrant and elegant, and it is a fact that women are particularly fond of Chiroubles—probably their favorite Beaujolais along with Fleurie, its granitic neighbor to the north. Chiroubles has a floral bouquet of iris and lily-of-the-valley, peony, violet, and faded roses, and fruity aromas ranging from red berries to cherry and prune.

FLEURIE: A WOMAN'S BEAUJOLAIS

As we move northward out of the Brouilly and Morgon districts, we leave behind a terrain of Carboniferous schist and lava to enter a land of pink granite, marking the final collapse of the Hercynian Mountains about 310 million years ago, when more magma squeezed through the stretched-out crust. Stalling and cooling below the surface and now uncovered by erosion, this pink granite outcrops throughout the vineyards. We already ran across it in the western hills of Régnié and Chiroubles, but north of Morgon it stretches across the entire piedmont zone, making up in particular the terroir of Fleurie.

Fleurie is an elegant, velvety, and very fragrant wine, competing with Chiroubles as the most feminine of Beaujolais Crus. It is perhaps no coincidence that the winemaking cooperative of the village was founded mostly by women, who were forced to take over the running of the vineyards during World War I when their husbands were sent to the front. Decades later, at the close of World War II, Marguerite Chabert (1899–1984) became the first woman to head a wine cooperative in France. She left a lasting mark on the vineyard and did much to put Fleurie on the map. At the cooperative today, several Fleurie cuvées are still entirely produced by women.

Fleurie wine has a carmine-red color, a floral bouquet of rose and violet, iris and peony—"fields of spring flowers," to quote oenologist Robert Joseph—and fruity aromas of peach and red berries. The village is quaint, with several cafés and restaurants (a rare commodity in the Beaujolais hills), and the vines climb up the slopes just behind it. The most spectacular of these hills is La Côte de la Madone ("The Madonna Hill"), crowned by a little chapel that offers a glorious view over the vineyard, the village, and the Saône valley (fig. 2.5). The vines cling to the granite bedrock, broken down into a thin cover of colorful pebbles that look very much like candy. The pebbles are made of pink, potassium-rich feldspar and milky quartz, quite a contrast with the dark blue and green amphiboles of Mont Brouilly in the south. These were different volcanic episodes, yielding strikingly different terroirs.

The Fleurie pink granite is calc-alkaline in nature—technically a monzonite—with interlocking crystals of potassium feldspar, calcium feldspar, quartz, and minor aluminum and iron-rich minerals, such as flakes of black mica. If such magma were to reach the surface rather than crystallize at depth, the resulting lava would be trachyandesite, like most of the lava that erupted at Mount Etna in Sicily or Mount Tambora in the Philippines. When such magma is rich in gas, it erupts violently, belching out great clouds of ash and pyroclastic flows.

There is no sign at Fleurie that magma erupted at the surface, and only erosion makes the granite outcrop at the surface today. High on the slopes, the rocky ground bears little to no soil and yields a very typical

Figure 2.5 The Madonna Hill (Côte de la Madone), crowned by its chapel, is one of the best "climates" of the Fleurie vineyard.

Beaujolais, with a fine mineral character. At the foot of the slope, near the village, the soil is deeper and richer in clay, yielding full-bodied, more complex wines. The very color of Fleurie changes from bright ruby when made from vines at the top of the hill to a darker hue obtained from vines at the bottom. On the vineyard's northern border, the wines are richer with deep structure, as the terroir segues into the neighboring and equally famous appellation of Moulin-à-Vent.

MOULIN-À-VENT

Known as the "King of Beaujolais," Moulin-à-Vent is perched on the pink granite hills above the village of Romanèche-Thorins. The vineyard earned its separate Cru appellation as early as 1936, but in fact the area's winemaking tradition goes back to Roman times. Quality wines produced by the rich villas of Romanesca, as the village was then named, are mentioned in writings dating to the third century AD. It is only much later, around 1850, that a windmill was erected on a pedestal of granite in the middle of the vineyard, and gave its name to the wine (fig. 2.6).

Figure 2.6 The vineyard of Moulin-à-Vent, with its famous windmill and the Beaujolais mountains on the horizon.

Moulin-à-Vent is famous for its strong tannic structure, and a fragrant bouquet of black currant and violet that evolves toward rose and iris, spices and forest undergrowth. The other characteristic of Moulin-à-Vent is that it ages remarkably well: up to ten years or even longer for the best bottles, like a true Burgundy. These properties, unusual for a Beaujolais, are often credited to the terroir's unique chemical composition, spiked with manganese and other rare metals.

Moulin-à-Vent bedrock is the same pink granite as at Fleurie, Régnié, and Chiroubles, but at the level of Romanèche-Thorins it is sliced up by the great fault system that runs along the hill front. Many of the cracks running through the granite were hydrothermal fissures brewing with hot, metal-rich fluids. They have now solidified into siliceous quartz veins that host concentrated amounts of commercially interesting metals, such as barium and manganese. The ore-rich veins brought prosperity to Romanèche-Thorins in the nineteenth century, when as many as four manganese mines were in operation on the outskirts of town.

Manganese has many uses. In steelmaking it is mixed with iron to produce strong alloys used in manufacturing bank vaults and plow blades; it is used as a cathodic metal in batteries and in coatings against rust and corrosion; and it serves as a violet coloring agent in glassmaking (in nature, manganese colors quartz violet, producing amethyst). In agriculture, manganese is an important trace element used in fertilizers, particularly for vegetables and citrus fruit. In the human body, about a dozen milligrams of manganese are stored in the bones, liver, and kidneys, and the element is an important ingredient of many types of enzymes, takes part in the body's absorption of vitamin B1, and neutralizes dangerous free radicals.

As for the effect of manganese on wine, we can only speculate. Maybe the manganese dispersed in the soil, by erosion and the leaching of the quartz veins, does play a role in the violet bouquet and special character of Moulin-à-Vent. More chemical analyses and wine tasting are needed to shed more light on the issue.

CHÉNAS

The smallest of the ten Crus of Beaujolais, with 265 hectares (662 acres) of vines, Chénas is the northern continuation of Moulin-à-Vent, a prestigious and somewhat greedy neighbor that has encroached on the southern part of Chénas territory to make more wine, given the growing demand for Moulin-à-Vent. In fact, the winemakers of Chénas are entitled to commercialize their wine under the name Moulin-à-Vent if they so desire. But winegrowers have their pride, so most Chénas is sold under its own name, and lives up to its neighbor.

Chénas is a wine of character, sharing with Moulin-à-Vent the same terroir of pink granite and manganese-rich quartz veins. It also has the same strong bouquet of violet. Otherwise, Chénas has a deep ruby color and complex aromas, both fruity and floral. On the fruit side, black currant and raspberry are dominant; on the flower side, besides violet, rose and peony stand out. Like Moulin-à-Vent, aging Chénas develops spicy and woody notes.

Chénas probably owes its name to the oak trees (*chêne*) that once covered the hill slope and were cut down by order of Capetian king Philip V in 1316. Three centuries later, in the days of the Musketeers, Chénas was the favorite wine of king Louis XIII (1601–1643).

JULIÉNAS AND SAINT-AMOUR

The town of Juliénas, probably named for Julius Caesar, is nestled in the Beaujolais hills at the center of a vast amphitheater open to the southeast, carved by the Mauvaise stream—a small confluent of the Saône. The vines occupy gentle slopes that get steeper as one approaches the footwalls of the amphitheater. The rock is predominantly pink granite—the northerly extension of the magmatic episode that occurred 320 million years ago. Here the granite is terminated by faulting, with a jump to schist and limestone on the down-dropped blocks leading down to the Saône valley, interspersed with pockets of sand and clay. This diversity in terroir, coupled with changing slope and orientation, confers to Juliénas a wide range of personalities and aromas.

Juliénas raised on granite is a light-bodied wine with a traditional bouquet of red berries leaning toward raspberry, and a touch of violet inherited from its neighbors Moulin-à-Vent and Chénas. Other lots are more reminiscent of Morgon Côte du Py, with gingerbread in their bouquet and quite a bit of structure—a character credited to the schisty nature of the soil. Finally, on the lower slopes, the clayey sands give more supple wines that are fruity and simple when young, but often become remarkably complex when they start aging, developing nuances of vanilla and spices, truffle and undergrowth.

To the northeast of Juliénas lies Saint-Amour: the tenth and last Cru of Beaujolais if traveling from the south, or the first if entering the region from the north. Such a romantic name as Saint-Amour calls to mind passionate lovers, or monks frolicking in the vineyards at harvesttime—but the reality is quite different.

The village owes its name to a Roman soldier named Amor who converted to Christianity and managed to escape the massacres perpetrated against Christians in Switzerland around AD 286, by co-Emperors Diocletian and Maximian. Amor fled across Rhône and Saône to the distant Beaujolais hills, where he founded a monastery—probably on the site of the present church—and was later elevated to sainthood.

With 310 hectares (775 acres) of vine, Saint-Amour is smaller but has even greater terroir variety than Juliénas, as it sits at the very transition between the granite and schists of Beaujolais and the younger limestone sediments of Burgundy. You can find at Saint-Amour some Triassic sandstone—a period we will explore in the next chapter—as well as schist and clay. There are a dozen different terroir "climates" at Saint-Amour, some named for the village church, such as Vers l'Église ("Toward Church") and Clos du Chapitre ("Lot of the Monastic Order"), others bearing more spiritual and passionate names, such as La Folie ("Folly") and En Paradis ("In Paradise").

Saint-Amour winemakers do not place great emphasis on small-scale variations of their terroir, stressing instead the overall character of their wine, of which they make two different versions. One is achieved by a process of short maceration—less than ten days in the vat—that brings out fruity aromas of raspberry and black currant, and occasionally peach and apricot. This Saint-Amour has a sparkling ruby color and is a light and mild wine, to be enjoyed young. It comes to maturity fifteen months or so after harvest, perfectly in time for Valentine's Day on February 14— its major marketing success. Nearly a quarter of the total production of Saint-Amour ships to the restaurants and wine cellars on that day, with its bottles bearing a romantic label featuring Cupid with bow and arrow.

The other style of Saint-Amour, which undergoes a longer maceration period, has more of a purple color. It is a wine worth keeping three or four years in the cellar, as it becomes more voluptuous with time, and develops extra aromas of spices and kirsch (cherry brandy).

Like all good Beaujolais wine, Saint-Amour goes very well with red meat and poultry, pâtés and cold cuts, cheeses, summertime picnics, and of course wedding banquets. The church of Saint-Amour came up with the clever idea of offering a wedding "confirmation" for married couples wishing to commemorate or renew their commitment. Along with actual weddings, these celebrations end up at the local restaurant, where regional dishes are served, and naturally washed down with the proper amount of Saint-Amour.

Here we end our rally across the ten Grands Crus of Beaujolais, on the border with Burgundy which features an entirely different terroir. The transition is displayed in the diverse building stones encountered in the walls of old houses and churches. As we proceed north into Burgundy, we see the pink granite progressively being replaced by red Triassic sandstone and golden Jurassic limestone. But before we explore Burgundy's "Jurassic Park" in the outskirts of Mâcon, the next chapter covers the Triassic period and takes us to its most representative terroir, outcropping on the slopes of the Alsatian rift valley.

FROM GRAPE TO WINE

The traditional way to make red wine is to pile grapes into an open vat and have them ferment one or two weeks. Most of this fermentation—when bacteria acts on the grape sugar, transforming it into alcohol—occurs during the first few days; the rest of the time is for extra maceration of the grape skins, increasing the wine's concentration in pigments, tannin, and aromatic molecules. After this phase, wine is transferred from the vat into oak barrels, where a second round of fermentation takes place, transforming the wine's malic acid into lactic acid. This process can take several months to over a year, after which the wine is bottled and sold.

Beaujolais winemaking is slightly different, which also contributes to its special flavor. Once the grapes are loaded in the vat, they are covered so that the oxygen supply is cut off. The bacteria then get to work, working their chemical transformation in the juice but also inside each individual grape (which are hand picked in the Beaujolais, so as not to bruise them); carbon dioxide builds up inside the vat. Under these reducing conditions, malic acid is transformed into ethanol, which considerably lowers the acidity. The wine acquires a typical aroma of red and black berries. This special technique of vinification performed on Beaujolais wine is called carbonic maceration. After this step is completed, the grape is pressed to extract the wine, which is then stored in vats or oak barrels for up to several months before bottling.

A new variation on the theme appeared in the 1990s, with a technique called hot prefermentation maceration (60 to 70°C [140 to 160°F]) that weakens the grape skin so that it prematurely releases color and aromas into the grape must. This type of Beaujolais is deep purple with violet streaks, and is recognizable by its powerful black-currant bouquet.

CHAPTER THREE

Alsace

Alsace holds a special place in the history of French vineyards: it is the northernmost region that produces wine, which it owes to a most fortunate climate, itself the result of a very special geological setting.

Alsace occupies a north–south-trending rift zone—a spectacular downdrop of the Earth's crust between the Vosges mountain range and the Black Forest—and its vineyard receives protection from rain-carrying western winds. The vineyard also faces east and the morning sun.

Known as the Fossé d'Alsace, the rift and its fault system shaped a remarkably complex terroir for the vineyard, sliding huge blocks of terrain downward along the fracture zone and juxtaposing rock layers that are so varied that winemakers naturally tried out a number of different grapes to derive the best wine from their land: Pinot Noir and Pinot Gris, Sylvaner and Riesling, Muscat and Gewurztraminer (see box).

The matching up of grape and terroir was not an easy task, and throughout history a number of varieties rose to prominence or sank into disgrace, under the influence of fashion, wars, and revolutions. Torn between France and Germany—both geologically and politically—and ravaged by two world wars, Alsace managed to preserve its vineyard and even propel it to new heights over the past decades, transcending its reputation as simply a varietal region and establishing the notion of terroir to develop a range of Grands Crus that today rival those of Bordeaux and Burgundy.

Alsace has taken full advantage of its geological and grape diversity. Burgundy and Bordeaux, as we shall later see, are characterized by a rather

THE VARIETALS OF ALSACE

Alsace wines are somewhat of an oddity in the classification of French appellations, in the sense that it is the grape variety that is listed prominently on the label, with the place of production playing a subordinate role (except for Grands Crus or renowned establishments, in which case the name of the place or that of the winemaker also commands large type). Here are the starring varieties of Alsace:

* **Pinot Noir** (9% of total wine production in Alsace): with a bouquet of black currant and raspberry, it accompanies cold cuts, cooked meats, and stew, and also produces a rosé wine.
* **Pinot Blanc** and its **Auxerrois** cousin (20% of the production): these yield supple wines with a light acidity that are a good match for seafood, sauerkraut, and quiche, and are also used in blends such as Edelzwicker.
* **Sylvaner** (12%): light and fruity, it accompanies seafood, sauerkraut, and is also used in blends.
* **Riesling** (23%): fine and racy, with a delicate bouquet, this variety is very sensitive and responsive to terroir, aromatically expressing its nuances. It accompanies fish and shellfish, poultry, veal, and sweetbread.
* **Gewurztraminer** (18%): this is a luscious wine, with a bouquet of fruit, roses, and oriental spices. It is a good match for foie gras, grilled lobster, and strong cheese like Munster, Roquefort, and other blue cheeses.
* **Pinot Gris** "Tokay" (12%): halfway between dry Riesling and sweet Gewurztraminer, Pinot Gris is a natural for poultry in sauce, scallops, and in the case of late-harvest batches, foie gras.
* **Muscat** (2%): drier than Gewurztraminer and Pinot Gris, this is a fruity varietal with a raisin aroma that is served as an aperitif or with dessert.

Alsace wines also include blends, which actually was the tradition in the region before varietals became the norm. The common blend is called **Edelzwicker** (1% of the Alsatian wine production) and mixes mostly Sylvaner and Pinot Blanc. The more upscale blend of Riesling, Pinot Gris, and Gewurztraminer is named **Gentil**.

limited range of soil and bedrock, whereas Alsace runs the whole gamut of granite and sand, salty evaporites, limestone and clay, to offer a surprisingly broad palette of aromas, best expressed in its white wines. Alsace is actually the only region in France to produce much more white wine than red (four times more, to be precise).

Alsatian white wine is a natural match for seafood, and it so happens that we find most of the ancestors of a good seafood platter in the local terroir, starting with oyster fossils. Before tectonic forces shaped the landscape into hills and rift valleys, Alsace was a flat land covered by a shallow sea, with marine mollusks and plankton dropping their shells and tests (outer skeletons) on its limy bottom.

These sediments illustrate a major chapter of France's geological history: the Triassic period that ran from 250 million to 200 million years ago. After the birth of France, which we witnessed in the terroir of Anjou, and the construction of a great mountain range, displayed in the Beaujolais hills, Alsace holds the best record of the destruction of these mountains, as collapse and erosion laid the land flat and a northern sea entered Europe through the lowlands of Holland and Germany.

These ancient marine sediments and their fossils could have been buried forever under younger sediment or milled down to dust by erosion, had it not been for a set of circumstances that preserved a few rare spots, in France and elsewhere, where they are displayed today at the surface.

The Rhine rift that slices through Alsace is such a privileged site. Because faults dropped down large sections of sedimentary strata behind a more elevated rift shoulder (the Vosges Massif), patches of Triassic sediment were protected by this high shoulder, which intercepted most of the rainfall before it could reach and erode them. For this reason, the piedmont hills of Alsace encapsulate one of the finest records of Triassic history worldwide, and some of the finest French vineyards as well.

A WINE TOUR OF ALSACE

One convenient way to visit the vineyards of Alsace is to follow the scenic wine route (La Route des Vins d'Alsace) that follows the strike of the rift valley, north to south, weaving up and down the vine-covered hills and connecting more than fifty townships over a distance of less than 200 kilometers (125 miles).

At its northern end, La Route des Vins d'Alsace starts half an hour west of Strasbourg at Marlenheim, a charming village at the foot of the hills. It is only fair that Marlenheim be the gateway to the Alsatian wine route, considering that its vineyard is cited as early as AD 589 in Gregory of Tours's chronicles: according to the Gallo-Roman historian, young Merovingian king Childebert II (570–596) owned vines in Marlenheim. The side story is that his governor, Count Droctulf, had plotted against him and was sentenced to forced labor in the royal vineyard, his head shaved and his ears cut off for good measure.

When Childebert's cousin Dagobert (602–639) inherited the Frankish kingdom, legend has it that he built two subterranean ducts to funnel wine—one for red, one for white—from the vineyard all the way to his palace in Strasbourg. The ducts do in fact exist, but they are Roman in origin and served to carry water from the foothills to the city. What Dagobert did do, however, was donate the best lots of the Marlenheim vineyard—the prized hill of Steinklotz—to the monastery of Haslach in AD 613. He was eleven years old at the time (and still a prince) but rather precocious, since he was said to have mounted horses and wielded heavy swords by the age of ten.

The Steinklotz hill, which Dagobert turned over to the knowledgeable custody of the monks, has since become one of the fifty-one Grands Crus of Alsace—the first Grand Cru that we encounter on our wine route. Starting at Marlenheim's town hall, you can follow a foot trail up into the Steinklotz vineyard that loops around the hill, which you can walk in about two hours (bring water and a hat, because there is no shade). Signs posted along the way indicate which varieties are grown where: Riesling, Pinot Gris, or Gewurztraminer. The soil and its substrate are marl and limestone. A small chapel on the hillslope marks the placement of one of the better lots, known to Steinklotz connoisseurs as the cuvée Bei der Kapelle ("By the Chapel" in German).

Leaving Marlenheim, as we drive down the Route des Vins d'Alsace we encounter a succession of medieval castles, fortified towns, and churches, among which are the town of Heiligenstein and its Landsberg castle, built from pink Vosges sandstone atop a granitic spur, on the slope of Mount Saint-Odile. The vineyard is famous for its rare grape variety named Klevener (elsewhere known as Traminer or Savagnin Rose) that was imported from northern Italy and planted here in the eighteenth century. Klevener does very well in this flint-rich terroir, from which it derives an earthy, smoky, and sometimes chalky bouquet, and a fruity, buttery aroma.

Farther south, near the city of Selestat, a valley perpendicular to the Alsatian rift penetrates westward into the Vosges mountain front. At its entrance, high on a spur, stands the spectacular Haut-Koenigsbourg castle that was built in the twelfth century. From its belvedere made of handsome pink sandstone and perched at an altitude of 800 meters (approx. 2,700 feet), we catch a panoramic view of the piedmonts and their vineyards, including Grand Cru Kintzheim; the plains of Alsace on the valley floor; and the opposite, eastern shoulder of the rift rising on the horizon and covered by Germany's Black Forest.

Still proceeding southward, we cross the picturesque villages of Ber-

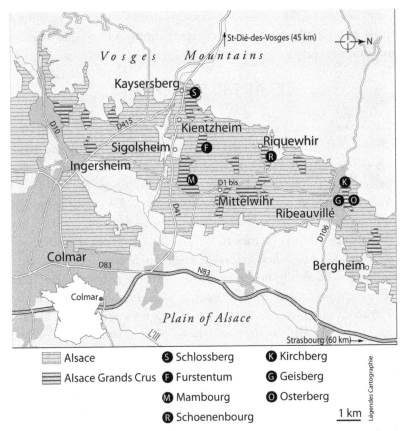

Figure 3.1 Map oriented west (north is to the right) across the plains of Alsace (*bottom*); the rift shoulder covered by vineyards (*center*); and the Vosges Mountains (*top*). Grands Crus mentioned in the text are labeled by the first letter of their name. (Map by Legendes Cartographie.)

heim, Ribeauvillé, and Riquewihr (fig. 3.1), each one the gateway to its own transversal valley, before reaching the larger Weiss River that digs back into the Vosges hills far enough to connect Alsace to Lorraine on the western side of the mountain range. It was a major passageway in Roman times, and still is today as route D 415, linking the Alsatian city of Colmar to the Lorraine city of Saint-Dié-des-Vosges.

It is at the mouth of this valley, fanning out into the vine-covered foothills, that we stop to take a closer look at the Alsace terroir, highlighted by three Grands Crus that line the mouth of the Weiss valley and lead into the Vosges Massif: the Mambourg Grand Cru over the village of Sigolsheim; Furstentum Grand Cru over the neighboring town of

Kientzheim; and the famous Schlossberg Grand Cru over the medieval village of Kaysersberg—Schlossberg being the first Grand Cru officially recognized in Alsace.

GRAND CRU SCHLOSSBERG

Halfway between the villages of Kientzheim and Kaysersberg, on the left side of the road, you can't miss noticing an elegant set of stone buildings—originally a seventeenth-century convent—surrounded by a garden of roses, a stone wall, and rows of vines. Alongside the wall runs a spring that gave its name to the estate: Weinbach ("Wine River" in German) (fig. 3.2).

The vineyards here go back to the Middle Ages. They are mentioned as early as AD 890, when Empress Richgard of Swabia (circa 843–circa 896), wife of Carolingian emperor Charles the Fat, donated them to the Abbey of Étival. Many centuries later, in 1612, the abbey passed on part of its land to Capuchin friars who erected the present buildings. Confiscated and sold off during the French Revolution, in 1898 the convent and its vineyards finally became the property of the Faller brothers, tanners and winemakers in the neighboring town of Kaysersberg. After World War II, Théo Faller (1911–1979) took over the estate from his father and uncle and worked hard to develop the vineyard and improve its quality. He lobbied for recognition of the Kaysersberg and other Alsatian terroirs as Grands Crus in their own right, on a par with those of other famous winegrowing regions in France.

Colette Faller—Théo's widow—and her two daughters Catherine and Laurence now preside over the destiny of the Weinbach estate, which has done so much for putting Alsatian terroir in the limelight. The domaine's 29 hectares (72½ acres) of vines include lots right outside the buildings (Clos des Capucins) and on the three hills overlooking the valley, each recognized today as a Grand Cru d'Alsace: Schlossberg, Furstentum, and Mambourg. Since the three hills belong to three different geological periods, a long story unfolds in the Riesling, Pinot Gris, and Gewurztraminer that the Weinbach ladies serve their guests and visitors. With a generous sense of hospitality, curious about everything that has to do with winemaking terroir and geology, they are expert oenologists as well and speak passionately of the bouquet and aroma of each one of their wines and terroirs.

Figure 3.2 The vineyards and small brook of the Weinbach ("Wine Brook") estate. In the background, the hills bearing two of its Grands Crus: Schlossberg (*left*) and the lower Furstentum (*right*).

Foremost among these is Grand Cru Schlossberg, which brings us back to the very foundations of the Alsatian terroir: the granites that constitute the crystalline basement of the area and were emplaced during the last gasps of the Hercynian period, 330 million years ago. Erosion wore down the mountain range, and much of its material was carried out by streams into the surrounding plains. The deep roots of the range were delivered of such a huge weight that they rose from the depths by rebound, a gravitational reaction known as isostasy.

During its slow ascent—a few centimeters per year—the drop in pressure experienced by the warm crustal rock forced it to partially melt. Some minerals melted completely before cooling down and crystallizing anew, whereas others only half melted, flowing in banded patterns like taffy. The end result is a rock mix that sometimes has the salt-and-pepper speckle of granite, sometimes the banded aspect of gneiss. The mass is shaped into a rounded hill, its slopes covered with vines and crowned by a forest. Its name is Schlossberg—"The Castle on the Hill"—because one of its rocky spurs serves as the pedestal for a medieval castle overlooking the town of Kaysersberg and its valley, passageway across the Vosges Mountains to the western province of Lorraine.

To the winemakers' delight, the Schlossberg vineyards are oriented due south. Much to their chagrin, the sunny slopes are steep, climbing from

230 meters to 350 meters (approx. 760 to approx. 1,160 feet) above sea level. Vine growers labored to pick rocks off the slope and build retaining walls, planting the vines in terraces. As for the granitic nature of the rock and soil, it does very well with Riesling, a grape that is particularly sensitive and aromatically responsive to different types of soil, though it probably reaches its best expression here. Pinot Gris and Gewurztraminer are also grown on the hill.

The soil on Schlossberg is composed of a thin layer of coarse sand, the result of the breaking down of granite. Rainwater is efficiently drained out by the steep slope and the fractured nature and porosity of the soil and bedrock, providing the hydraulic stress necessary to make great wine.

Schlossberg's granitic terroir also brings precious chemical supplements to the vine. Broken down into its constituent minerals, granite yields black mica rich in iron and magnesium, and feldspar turning into a clay rich in potassium and phosphorus. As for the quartz grains, they are chemically inert but play a positive physical role by keeping the soil granular and well aerated. Under these conditions, Riesling gives a sharp and mineral wine with a bouquet that can be both floral and fruity, often with a citrus touch and sufficient acidity to help it age well.

What is surprising—and also reveals the magic touch of terroir—is that the wine's characteristics vary significantly along the slope, a function of soil and microclimate. At the top of the hill, the soil is thin and the wine fine and taut. At midslope, the vines enjoy warmer temperatures and the soil is thicker. Older vines are encouraged here, producing the best batches or cuvées, which are characterized by mature aromas and a rich body. Last, at the bottom of the hill, where the soil is deepest and the breakdown of feldspar produces much fine clay (hence a "cool" soil), the grapes can be harvested later and yield a wine that is particularly full bodied and rich.

Translated into gastronomic terms, this means that one should serve the taut Riesling of the hillcrest with grilled fish, oysters, and seafood; the grand cuvée of the midslope with lobster and fine fish; and the richest Riesling of the lower slope with fish in sauce. All this on only one hill!

GEWURZTRAMINER: LIMESTONE STRIKES BACK

Schlossberg's granitic hill lines up with Furstentum and Mambourg, two other steep hills jutting into the rift valley along the Weiss River, with vineyards facing south and southeast. Here we are no longer in granitic terrain, because the little valley that separates Schlossberg from Furstentum conceals a major tectonic break: the great Vosgian fault line.

As stated earlier, the piedmont hills along the Vosges Mountains' front were formed by great blocks sliding down along steep faults during the opening of the Rhine rift zone—a several-hundred-meter drop that placed younger sedimentary rock on the same level as the ancient granites. Crossing the little valley between Schlossberg and Furstentum, we jump 100 million years from the roots of an old mountain range to the marine sediments that later covered the crustal basement.

Sediments outcropping at Furstentum through a thin, brown, and pebbly soil are limestone, marl, and sandstone from the Middle Jurassic, aged about 175 million years. They tell the story of a warm sea that came and went several times across the land, dropping limestone when it was at its deepest, and sandstone when it was shallow and turned into alluvial plains where sand and gravel collected.

Furstentum's "layer cake" of sediments yields just as sophisticated a Riesling as Schlossberg granite does, but a Riesling that needs more time—ideally seven to eight years—to develop its full bouquet. It is a little less mineral and more opulent than its counterpart at Schlossberg. Furstentum's marly, calcareous, and sandy terroir is also well suited to the other two top grape varieties of Alsace: Pinot Gris and Gewurztraminer. In fact, wine reviewers often cite the Furstentum hill as the best Gewurztraminer terroir in Alsace.

Gewurztraminer is a grape variety derived from Savagnin Rose, itself originating from the province of Tyrol in the Central Alps. It reaches its best expression in Alsace, where it represents 20 percent of the total vineyard. Gewurztraminer yields a sumptuous wine, sweet and mellow, with aromas of candied citrus fruit and oriental spices, rose, and jasmine. Besides its great solo performance as an aperitif and its faithful allegiance to foie gras, Gewurztraminer goes remarkably well with exotic dishes, such as tajine or curry, and with strong-flavored cheese, like the Alsatian Munster that originates from the next valley over, 2 kilometers (1¼ miles) south of the vineyards.

After Schlossberg and Furstentum, the third, easternmost hill of the piedmont alignment is Mambourg, advanced enough in the rift valley to catch the sun and stay out of the shadow of the Vosges crest line until late in the afternoon, maximizing the number of sun hours for its vines.

Gewurztraminer, Pinot Gris "Tokay," and Muscat all love this extra dose of sunshine procured by the hill's southeastern exposure, as well as the soil made up of both limestone and marl, well balanced in minerals.

This move from Furstentum to Mambourg is another time travel across a fault zone, which has us jump from the Middle Jurassic, 170 million years ago, into the middle of the Cenozoic—the era of mammals—less than 50 million years ago. It is around this time that the uplift, probably thermally driven, and the extension of the Earth's crust in the east of France initiated the foundering of the central part that became the Alsatian rift, and the uplift of the encasing rift shoulders. Torrents washed down the slopes, plucking, mixing, and recycling the strata encountered—older sandstone, marl, and limestone—into new layers of sand, gravel, and cobbles several hundred meters thick.

RIBEAUVILLÉ: THE SEA COMES TO ALSACE

By crossing three adjacent vineyards over a distance of 3 kilometers (2 miles), from the western hill of Schlossberg to the eastern hill of Mambourg, we thus clocked 300 million years in time, from the days of the mighty Hercynian mountain range to the flattening of the land and finally to the blistering of the crust and the opening up of the Rhine rift valley.

The faulted contact zone between the down-dropped valley floor and the rift shoulder is extremely complex. In the area just described, at least a dozen parallel faults slice up the crust, one every few hundred meters over a total width of 4 to 5 kilometers (2½ to 3 miles) down the slope and into the valley, defining a set of terraces. These major north–south-trending faults are crosscut by small perpendicular faults striking east to west, so that the wide fault zone is cut up into a giant mosaic or chessboard. Because of fault motion and the whims of erosion, adjacent squares of the board tend to boast different rock strata and form different terroirs (fig. 3.3).

We have seen at Schlossberg and Mambourg extremes in ages, with respectively some of the oldest and some of the youngest bedrock strata in the region. If we move several squares north on the giant "fault board," we discover other terroirs that have made Alsace geologically famous: the rare, beautiful strata of the Triassic period.

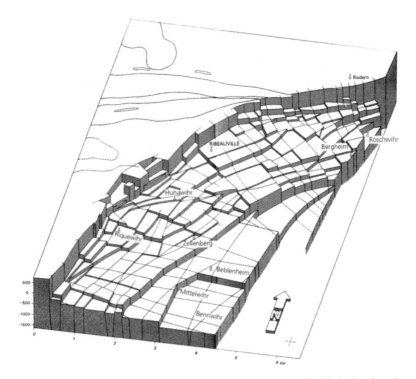

Figure 3.3 The Ribeauvillé fault zone slices up the edge of the Alsatian rift into a mosaic of fault blocks, down-dropped to different elevations. The "squares" often display bedrock of different ages at the surface, which explains the variety of Alsatian terroir. (Diagram from the thesis of Georges Hirlemann, 1970.)

The Triassic period ranges from 250 million to 200 million years ago. At the time, all continents were united in a single block—a supercontinent named Pangaea by geologists—and the climate was hot and dry. The world had just emerged from a major mass extinction (the End Permian event), and the surviving reptiles diversified remarkably well during that period. They became the ruling vertebrates of the ecosystem and developed, among other major successes, a new reptile group that would become known as the dinosaurs.

This was also the time when the Hercynian Mountains of Europe finished collapsing and eroding away. The only elevations still remaining on the land surface of Triassic France were a couple of granitic massifs in the center and the west of the country (Massif Central and Massif Armoricain; fig. 3.4), which had streams running down their slopes and converging east. Alluvial fans with multiple channels covered what would later

Figure 3.4 France during the Triassic period, 250 million years ago. The eastern edge of the country is penetrated by a shallow Germanic Sea that stacked up on its bottom the gypsum, limestone, and marl of future Alsatian terroirs. (Map by Pierre-Emmanuel Paulis.)

become Alsace. Their sand and gravel are now consolidated into multicolored sandstone, a hard rock used for building castles and ramparts and also for making mills for grinding grain. In the finer sand can be found fossils of small terrestrial vertebrates, and in particular some nice scorpion specimens, as well as fossil branches and even entire trunks of conifers—the dominant type of tree at the time.

After this rather arid and gravelly debut, the Triassic period in France took on a completely different dimension with the invasion of the sea. Great tectonic movements were brewing south and west of the future country as the supercontinent of Pangaea began to split. Seafloor spreading and the growth of volcanic rifts underwater drove a worldwide sea-level rise, and marine waters flowed into the continental lowlands. In France this invasion came from the east through Germany, and what were once deltas and alluvial plains in Alsace were submerged under dozens of meters of water, welcoming a brave new world of plankton, mollusks, and other invertebrates. Upon dying, these creatures contributed their shells to building up one of the major limestone strata of the Alsatian terroir.

The calcareous sediments of this Triassic Germanic Sea, deposited 240 million to 230 million years ago, are named *Muschelkalk*—German for "shell limestone." They are represented on many squares of the Alsatian chessboard, one being the township of Ribeauvillé.

Set at the entrance of another trans-Vosgian pass—the Strengbach valley—Ribeauvillé (fig. 3.5) is a peaceful, charming village, famous for its three thirteenth-century castles that stand on rocky spurs of granite: Saint-Ulrich; Girsberg, which faces Saint-Ulrich across a gully; and Haut-Ribeaupierre, which overlooks the other two from the highest hillcrest.

Below the castles and the wood-covered upper slopes, vines cover the hill front known as the Hagel, a granitic terroir that brings to Riesling a classic mineral character. Downstream from a major hidden fault, the down-dropped *Muschelkalk* limestone makes up the terroir of the next three hills standing over the village. All three are Grands Crus of Alsace: Kirchberg ("Church Hill") upstream, Geisberg ("Grazing Hill") in the middle, and Osterberg ("Eastern Hill") downstream.

The exceptional quality of the wine from Kirchberg was mentioned in

Figure 3.5 Medieval castles (namely Saint-Ulrich, *left*) tower over the village of Ribeauvillé (*right*). The central vine-covered slope with the white building is the granitic Hagel terroir.

writing as early as 1328. Under a pebbly soil only centimeters thick lies the *Muschelkalk*: the very first bench of limestone deposited by the Germanic Sea. The water was shallow at the time; rivers mixed in loads of sand and clay; and episodes of evaporation precipitated layers of salt as well. Rather than call the *Muschelkalk* a simple limestone, geologists describe it as "sandy, gypsiferous marl." Grapes derive remarkable aromas from this varied, finely mixed terroir, so much so that oenologists can usually tell a Grand Cru Kirchberg blindfolded: a full-bodied wine with a fine bouquet of spicy notes, leaning toward anise, and often the sweet touch of noble rot brought on by the grapes' late harvest in October, when they are affected by patches of mist rising up from the valley.

A few hundred meters to the east, Geisberg and Osterberg have a similar *Muschelkalk* foundation, with limestone and marl at the bottom and younger, more colorful marl near the summit, covered by a stony, clay-rich soil. These two Grands Crus also produce top Riesling wines, distinctly mineral, as well as some Gewurztraminer and Pinot Gris.

To proceed higher up the stratigraphic ladder of the Triassic period, we need to move several squares to the south on our Alsatian "fault board" and visit the neighboring village of Riquewihr.

RIQUEWIHR: THE AGE OF SALT

Riquewihr is one of the most typical, picturesque towns of Alsace, and was spared from the bombs of World War II. It features a fortified city wall built of Vosgian sandstone, dating to the thirteenth century; three churches; paved, narrow streets; and old, half-timbered houses with wooden balconies and peaked tile roofs.

The south-facing hillslope just north of the village is covered with vines that extend to the neighboring town of Zellenberg, one kilometer (six-tenths of a mile) downstream. This hill is named Schœnenbourg—another Grand Cru of Alsace, whose reputation reaches back to 1644, when Swiss cartographer Matthaüs Merian described it as "the most exquisite wine in the country." Mostly Riesling is grown here, as well as some Pinot Gris and Muscat, all known to develop rich and powerful aromas that come not from *Muschelkalk* but from the next strata of sediments that follow it in the Triassic period: the Keuper marls.

Around 230 million years ago, the Germanic Sea began to withdraw from Alsace. As the sea level dropped, deepwater limestone and marl were replaced by shallow-water reefs, then by a new set of marls more typical of coastal lagoons and swamps. Under hot and arid climes, the ebbing sea left behind standing bodies of salt water that precipitated gypsum and chloride salts as they evaporated.

These mixed marl and salt layers reaching dozens of meters in thickness make up the Schœnenbourg hill. To view a great cross section of the sequence, one need only cross the city wall through its northern gate. The passageway leads to a parking lot; beyond it, a quarry wall cuts off the base of the hill just below the vineyard. Here, gypsum was quarried centuries ago for use in making plaster.

In this rock face, gypsum and marl alternate in thin layers. The clay-rich marl is gray to bluish, and the gypsum creamy pink to rust-colored, with a fibrous texture and a mother-of-pearl sheen. Geologists aptly named this sequence the iridescent marls (*marnes irisées*).

Digging their roots into this clayey, salty substrate, the Riesling vines of Grand Cru Schœnenbourg combine mineral and fruity aromas, leaning toward blood orange, along with the rich, full-bodied taste provided by the marly terroir. Winemaker Jean-Michel Deiss goes even further, spotting in Schœnenbourg Riesling "whiffs of iodine, of seaweed, of beaches at low tide." Could it be the salts present in the soil that transmit to the wine this ghostly marine impression?

By the end of the Triassic, the Germanic Sea completely withdrew from Alsace. Besides the excellent substrate for the vineyard that it left behind along with the salt that was later mined in Alsace and Lorraine, the retreating sea left in the Keuper iridescent marls a few hints of the continental life-forms that roamed the coastline a little over 200 million years ago. Paleontologists have found footprints in the old, dried-up mud-flats—tracks that belong to large amphibians and reptiles—while the fine sandstones of floodplain river channels fossilized the bones of new lines of animals that were moving into the terrestrial ecosystem. One of the interesting species discovered in the Late Triassic marls is a primitive mammal named *Morganucodon*, which was the size and shape of a shrew.

Mammals were to remain small for another 100 million years, but a competing line of reptiles also made its debut in the Triassic and met with great success. We find the fossilized bones of one of its early prototypes, *Plateosaurus*, in Germany and in the French Jura Mountains: a reptile

7 to 8 meters long (20 to 25 feet) with a small head mounted on a long neck, and an equally long balancing tail. *Plateosaurus* is one of the first species of dinosaur, paving the way for an army of giant quadrupeds and bipeds that would follow during the Jurassic and Cretaceous periods.

Before the dinosaurs rose to power, a new crisis of the biosphere occurred at the end of the Triassic, precisely 201.5 million years ago. Once again, the tree of life was violently shaken, and more than half the animal species disappeared on land and in the seas. Asteroid impact, catastrophic volcanic eruptions, or some other cause? The jury is still out. All we know is that enough survivors made it through to launch a new wave of evolution in the Jurassic period, and establish for good the reign of the dinosaurs.

RIESLING SYMPHONY

In Alsace, Riesling is the grape variety that best translates the nuances of terroir into the wine. Etienne Loew, an up-and-coming winemaker from Westhoffen, 25 kilometers (15 miles) west of Strasbourg, makes a point of distinguishing his lots of Riesling grown on different terroirs—he produces a different wine for each hill and rock type.

Westhoffen is located in a wide fault zone at the foot of the Vosges Mountains, with a down-dropped block known as the Balbronn Graben framed by two uplifted shoulders: the Kronthal hills to the north and the Dangolsheim hills to the south. In the down-dropped block, partially protected from erosion, sedimentary layers from several geological periods have survived, separated by faults, and each constitutes a distinct terroir. Over the years, Etienne Loew has managed to purchase lots on these different terroirs and grows several grape types on each; but Riesling stands out as the one grape that most clearly translates into its wine the differences in soil, and plays out an aromatic score—a true Riesling symphony.

At the bottom of the geological "octave" stands the oldest key: *Buntsandstein* ("stone made of colored sand" in German). This sandstone-based, silica-rich terroir above Westhoffen was already well known to the Romans before it was taken over by Friars in the Middle Ages, earning it the name of Bruderbach ("Stream of the Brothers"). The south-facing slope takes advantage of sandstone and limestone scree to yield a Riesling wine with a citrus bouquet leaning toward grapefruit, along with a fine acidity provided by the limestone and aromas that can range from bitter orange to peppermint.

Ten million years later in our terroir octave, we have *Muschelkalk* limestone brought in by the marine invasion of the Middle Triassic period. Near Westhoffen, Etienne Loew possesses lots on *Muschelkalk* slopes facing west; these yield a sound, pure Riesling with a mineral and floral bouquet, lingering acidity, and a flint-stone smoky touch.

The third note in our symphony, the Ostenberg ("East Hill") terroir straddles a fault and juxtaposes two limestones from two different ages—the gray *Muschelkalk* and a more recent yellow limestone—along with green marl. The result is a Riesling that is both fresh and lively.

The fourth note is played by the Sussenberg ("Sweet Hill") terroir, on steep slopes facing south and made of young yellow limestone that has broken up into cobbles. Riesling here yields a powerful wine with a bouquet of hazelnut and dried fruit, and a long and balanced palate spiked by a puckering sensation provided by the limestone.

The final note is Grand Cru Altenberg-de-Bergbieten, a hill 3 kilometers (2 miles) south of Westhoffen, where the terroir is a gypsum-rich black marl. This last Riesling offered by Etienne Loew has a fine acidic minerality, tart bonbon flavor, and aromas of black currant, citrus fruit, and rhubarb.

CHAPTER FOUR

Pouilly-Fuissé and Other Wines of Mâconnais

At the southern tip of Burgundy, before we reach the Beaujolais hills, the landscape ramps up like a wave toward the west and forms a crest of towering cliffs that hang over the vineyards. This wild and beautiful region is known as the Mâconnais, and its starring wine is Pouilly-Fuissé.

Pouilly-Fuissé is a dry, fruity white wine that rose to stardom in the 1970s and made a particular splash in the United States, mostly because it is a Chardonnay—the most popular varietal in America—and also because Pouilly-Fuissé (poo-yee fwee-SAY) is so awkward and fun to pronounce that it gets widespread name recognition.

To the French, the wine is well known, but the landscape is even more famous, because several promontories along the cliff face host archeological treasures. La Roche ("The Rock") de Solutré in particular has yielded a wealth of arrowheads and carving tools dating to the Stone Age (twenty-two thousand years before present), when *Homo sapiens* ambushed reindeer and wild horses in the area. Much more recently, the rock of Solutré was a rallying point for resistance fighters during World War II. After the war, a group of veterans of the underground movement met at Solutré every year to commemorate their brotherhood, including freedom fighter and later French president François Mitterand (1916–1996). His yearly hike to the top of the cliff, surrounded by his closest allies and cabinet members, was well covered by the media.

The hike up the back of the Solutré hill is well worth the effort (fig. 4.1). One can imagine what a great vantage point it offered Paleolithic hunters for spotting herds of horses and reindeer and planning their

SOLUTRÉ IN THE STONE AGE

At the foot of the Solutré rock, vine growers had long noticed that a thick bed of animal bones—mostly horse skeletons—was buried close to the surface. They would collect the bone fragments and sprinkle them in the vineyard to add phosphate to the soil. This astonishing wealth of animal bones eventually caught the attention of scientists in the mid-nineteenth century, namely geologists and archeologists Adrien Arcelin and Henry Testot-Ferry, who in 1866 conducted the first digs on the site of Cros du Charnier ("The Bone Hole").

The bone layer extends across approximately one hectare (roughly 2.5 acres) and is over one meter (3 feet) thick in some places, which amounts to about one hundred thousand horses slain or butchered on the site. The two scientists discovered stone tools and arrowheads exquisitely chiseled in the shape of laurel leaves, and made the connection with very old human skeletons discovered around the same time (1868) in Dordogne, at another *Cros* (or "hole") known as Cro-Magnon.

No human skeletons were discovered at Solutré, which suggests that the site was more of a hunting center or a meat-carving workshop than permanent living quarters. The site also happens to have been occupied many times in prehistory, from the time *Homo sapiens* (Cro-Magnon man) settled in France 35,000 years ago, and especially from 22,000 to 18,000 years before present, during the coldest period of the most recent Ice Age. It is during those four millennia that the hunter-gatherers developed the art of toolmaking and carved from flint the laurel-shaped arrowheads found at Solutré—a Paleolithic period and industry aptly named the Solutrean period.

Ice caps had reached their greatest extension at the time, down to the southern coast of the British Isles. Cold winds blew across a then dried-out English Channel and onto the northern plains of France, which resembled today's icy tundra of Siberia. The small community of hunters had retreated southward, and the limestone cliffs of Solutré were probably their northernmost hunting posts. They hunted horses and reindeer, as well as aurochs, wooly rhinoceroses, and mammoths.

An archeological museum along the trail (200 meters [600 feet] from the parking lot), at the foot of the cliff of Solutré, exhibits Solutrean arrowheads and stone tools, and presents a history of the digs. Opening hours and practical information can be obtained at 03 85 35 85 24.

hunting tactics. Today, the landscape is perfectly still and peaceful. To the east beyond the foothills flows the Saône River, and on a clear day you can make out the Alps and the snow-capped Mont Blanc on the distant horizon. Looking north and south, the view follows the crest of the Mâconnais hills: 2 kilometers (1¼ miles) to the north, the twin promontory of Vergisson mirrors the rock of Solutré; 2 kilometers to the south rises the wooded hill of Mont Pouilly. Finally, to the west, a rolling valley blanketed with vines surrounds the small village of Solutré, backed up by a parallel line of hills that protect the vineyard from the rain-carrying

Figure 4.1 View to the north from the top of Solutré rock. The Pouilly-Fuissé vineyard surrounds the village of Vergisson, with the towering Jurassic cuesta in the background.

western winds. Altogether, the panorama is breathtaking, and as President Mitterand wrote in his book *La Paille et le grain*, "From here I can see what moves in, what moves out, and above all what stands still."

A BIT OF HISTORY

From the crest of the hill, the Pouilly-Fuissé vineyard resembles a giant, malleable checkerboard that Salvador Dali or some other surrealist artist painted over the rolling landscape. But this particular checkerboard has only white squares, since the exclusive grape variety here is Chardonnay. Over the years, its competitors, Pinot Noir and Gamay, were pushed down into the Saône valley, where the production of red wine is concentrated today (appellations Mâcon and Mâcon-Villages).

Chardonnay grapes now occupy the best terroir, but it wasn't always so. In centuries past, red wine had more success than white, a reputation that can be traced to Roman times.

The first vines were planted around what is now the town of Mâcon by Celtic tribes during the second century BC. When the region was overrun by Julius Caesar and his army, one of his generals, Fussiacus, set up camp at the foot of a hill, from which flowed a precious spring. A village developed on-site and took the name of its founder—Fussiacus becoming Fuissé. As for the neighboring town of Pouilly, the origin of its name is less clear. It might derive from another Roman soldier's name—Paulium, for Paul—or originate instead from the Celtic word *poul*, meaning "pond" or "swamp," although there is no standing body of water left at Pouilly today.

Be that as it may, the need for water in the foothills was soon supplanted by the need for wine. By the beginning of the Middle Ages, vines had spread over the sun-facing slopes, initially at the lower elevations. Vine growers were faced with rocky ground littered with limestone scree. It took a great deal of hard labor, and careful management by the important Cistercian Abbey of Cluny—founded in AD 909 some 20 kilometers (approx. 12 miles) to the northwest—for the local winemaking industry to truly take off. Endowed with the power and responsibility to make the land flourish, the monks of Cluny took great care in selecting and crossing the grape varieties, choosing those best adapted to the soil and climate, figuring out how best to prune the branches, and perfecting the art of winemaking itself. The influence of the Cluny order is apparent in the many religious monuments erected in the area during the tenth century, starting with the elegant Romanesque church of Solutré and a smaller church at Fuissé, a few steps from the village-founding spring. In the western hills stands the chapel of Grange du Bois, overlooking the vineyard and keeping watch over its own spring: *la fontaine aux Ladres*. Its water is said to cure skin diseases and welcomed many a procession of the devout in the past, although local wine-tasting cellars have more of a following today.

With their monopoly over wine and water, the Cistercian monks ruled supreme throughout the Middle Ages. Among their many accomplishments, they developed the quality grape variety known as Chardonnay: a cross between Pinot Noir and an ancient grape named Gouais. Today, Chardonnay is the exclusive grape of the Pouilly-Fuissé appellation and of all top-notch Burgundy whites.

The Cistercians had no trouble filling their cellars and supplying the local nobility with great wine, but the Mâconnais wine industry had a rougher time penetrating the national market, faced as it was with formidable competition from northern Burgundy, Champagne, and the Loire valley. As bishops and dukes progressively took over the vineyards from the monks, they increased the exportation of Mâcon wine by ship toward the south, by way of the navigable Saône and Rhône Rivers. They still had trouble reaching Paris, for that entailed carting the barrels over the Mâconnais Mountains, then loading barges on the Yonne River, sailing downstream to where it joined the Seine, and proceeding to the capital.

According to legend, the Mâconnais wines got their break from the marketing tenacity and savvy of Claude Brosse (1656–1731), a winemaker from Charnay-lès-Mâcon, a village near Fuissé. Around the year 1700 — no one knows the exact date — Brosse loaded half a dozen barrels of wine onto his ox-driven cart and headed for Versailles to try selling them to the court of Louis XIV. Brosse supposedly bribed the guards to get in to the palace courtyard, then joined the congregation in the royal chapel, where Mass was being held in the presence of the Sun King himself.

It so happens that Claude Brosse was a very tall fellow: over 6 feet (close to 2 meters), when the average height at the time was closer to 5 feet (1.60 meters). As the story goes, when communion was given and the worshippers were asked to kneel, the king noticed Brosse in the back row, since his head stuck out above everyone else's. Believing the man was still standing, Louis sent a guard to force him to his knees. Upon learning that Brosse *was* kneeling, the king was so surprised that he arranged to meet the man after Mass. Brosse naturally jumped at the chance to show off his wine, and Louis found it quite good. Seeing this, the royal courtiers promptly raved over it as well and ordered barrels from the newcomer, making Claude Brosse a rich man and putting Mâcon wine on the map.

From the 1700s onward, Mâconnais wine had its ups and downs, depending on the competition from rival regions and fluctuating fashions and markets. Surges in output often led to a drop in quality, and the habit of blending wines from many different lots did nothing to improve them. Some of the production was even funneled into making low-grade sparkling wine.

Mâcon wines pulled out of their nosedive during the twentieth century, when conscientious winemakers decided to curb the production and make the vines struggle for water and thereby concentrate aroma, raising their sugar content in the process. The resulting wines became stronger and more complex, and even reached excellence in the most favorable terroirs — those crowned in 1936 by the now-famous appellation of Pouilly-Fuissé.

THE SUN KING–AND WINE CONNOISSEUR–LOUIS XIV

Louis XIV (1638–1715) was an eclectic king when it came to wine, and the hearty Mâcon reds that Claude Brosse brought to Versailles were only one of a long list of his favorites. He enjoyed the Chenin Blanc of the Savennières vineyards in the Loire valley, as we saw in chapter 1. He also fancied red wines from Bordeaux (namely Haut-Brion) and those from the southern vineyards of Languedoc (namely Fitou), as well as the flavorful rosé wines of Tavel in the Rhône valley.

But Louis XIV drank mostly Burgundy and Champagne wines. He took a particular liking to Côte de Beaune Burgundy, especially Volnay, which he found *souple et salutaire* ("supple and healthy"). The Côte de Nuits Burgundies were also among his favorites, as far north as Irancy.

The Champagne fancied by the Sun King was not a sparkling wine until the end of his reign, when Dom Pérignon mastered the wine's nasty tendency, once in a bottle, to ferment a second time and spawn gas and bubbles. Before the rise and fame of sparkling Champagne at the royal court, which actually happened during the reign of his son Louis XV, Louis XIV much preferred still Champagne wines, such as the red Bouzy (served at his crowning at the Reims cathedral in June of 1654) and Rosé des Riceys.

The Sun King's enjoyment of both Burgundy and Champagne wines sparked a fierce rivalry between the two regions that was backed by their scholars and doctors, for wine was the most important medicine of the time. The king's chief doctor, Guy-Crescent Fagon, prescribed Burgundy red in order to curb his frequent attacks of gout (with a few drops of cinchona alcohol added for good measure to combat fever). Besides, the Paris school of medicine had long ruled that "Beaune wine is the healthiest and most agreeable beverage." The Champagne faculty of Reims fought back with a publication, "The Superiority in taste and healthiness of Champagne over Burgundy wine," and the feud went on for years.

No wonder, then, that Louis XIV ended up commanding that an official classification of French wines be developed, to bring his country's delectable assets into clearer focus.

A CAR TRIP THROUGH POUILLY-FUISSÉ

The Pouilly-Fuissé appellation stretches across five townships hugging the piedmont hills and escarpments of the Mâconnais range: Vergisson, Solutré, Pouilly, Fuissé, and Chaintré (from north to south). Over a distance of no more than 5 kilometers (3 miles), the range of aromas is striking, as you can discover for yourself by visiting the Atrium, a winemakers' association located next to the Solutré church. Here you will be treated to a wine-tasting comparison of the five terroirs corresponding to each township.

If you wish to pursue your exploration in detail, the Atrium sells over one hundred different Pouilly-Fuissés representing nearly half the winemakers in the area. Altogether, the appellation hosts 250 estates sharing 700 hectares (280 acres) of vineyards (fig. 4.2). Add to this constellation

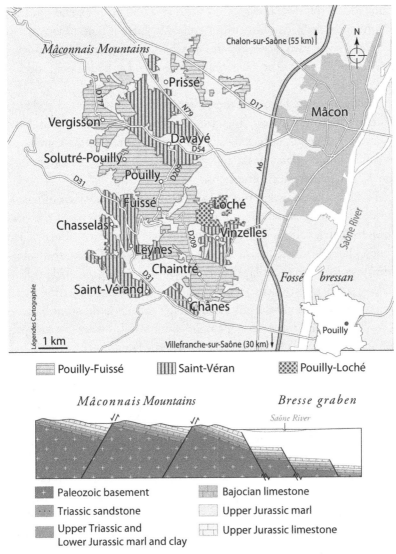

Figure 4.2a, b *Top*, the Pouilly-Fuissé AOC vineyard (and the satellite appellations of Saint-Véran and Pouilly-Loché). (Map by Legendes Cartographie.) *Bottom*, cross section of the area showing the down-dropping and tilting of major blocks between parallel faults, with often an escarpment or a cuesta marking the location of the fault.

of winemakers the area's diversity of geological strata, slope, and orientation, and you will easily understand why there is such a range of Pouilly-Fuissés—and probably more than one that suits your taste!

If we enter the area from the south and the township of Chaintré, we are leaving behind the Beaujolais region and its hills of granite and volcanic rock. However, a ridge of granite and lava actually extends into the southern border of Pouilly-Fuissé, flanked by Jurassic limestone. Most of Chaintré's vineyard faces east, and its Chardonnay grapes ripen quickly in the morning sun. The wines here are described as fruity, rich, and full bodied.

Following route D 209, we then enter the township of Fuissé, still hugging the volcanic hillcrest. The ground is covered with small chunks of rhyolite: a siliceous, taffy-like lava that is colored red, apricot, green, or even black in its glassy obsidian form, emplaced over 300 million years ago by gas-rich explosive eruptions.

Past the volcanic ridge, route D 209 dives toward the village of Fuissé, nestled below an amphitheater filled with vines. At this point the road folds back in a hairpin through Carboniferous coal-rich sediment, then follows Jurassic limestone down to the bottom. A small river, the Romanin, flows between the village's homes—the very spring that had inspired Caesar's officer Fussiacus to establish his camp on-site. At the point where it springs out of the hillside, the water collects in the stone basin of a washhouse, and the steep slope right above it holds vines in a lot aptly named Vignes sur la Fontaine ("Vines over the Fountain").

Past the spring, the washhouse, and a tenth-century Romanesque chapel, the village stretches out along the vine-covered hillside. Climbing the left, western side, the lots reach the tree line and bear evocative names, such as Les Châtaigniers ("The Chestnut Trees"), Les Vignes Blanches ("The White Vines"), and Les Chardonnets—an old spelling for the Chardonnay grape, and probably one of the first lots to feature it.

On the right-hand side of the road, the amphitheater levels out and stretches toward the Saône valley as a broad plateau known as Beauregard. Established on the youngest sedimentary strata of the Jurassic period, the plateau mixes coarse limestone and finer, clay-rich marl. Wines from this mixed terroir—those from the lots Vers Cras and La Croix Pardon, for instance—are known for their remarkable balance, mellowness, and finesse. As a whole, the wines of Fuissé are said to be the most balanced, elegant, and complex of the entire Pouilly-Fuissé appellation.

As the road leaves Fuissé, it enters the township of Pouilly, where the terroir is roughly the same. Lots are named Aux Cailloux ("Where the Stones Are") or even simply Pouilly, and produce wine that experts qualify as being particularly rich bodied, yet fine.

The road then follows a saddle between two hills before taking a bend to the west, running alongside the rising ramp of the Solutré rock until reaching its namesake village at the foot of the cliff. Surrounding the base of the promontory, in horseshoe fashion, lie the oldest Jurassic strata. The lots closest to the village have names such as Les Brûlés ("The Burned Ones"), Le Clos ("The Walled Lot"), or Vignes Derrière ("Vines at the Back"), whereas the lot in front of the cliff is named Vers la Roche ("Toward the Rock").

The wines from Solutré are finer and more mineral than those from the more southerly terroirs we just crossed, and the trend continues as we drive on toward Vergisson, the northernmost township of the appellation. We first pass in front of the prow of the Solutré rock—a photo opportunity stop—then climb over a vine-covered hillock to finally plunge toward the deepest, oldest strata of the Jurassic landscape.

Now the road goes through a dark patch of woods as it travels down to the bottom of a deep hollow, then climbs back into the sunlight past a row of houses and wine cellars overlooking the village of Vergisson, with the limestone cliff as a backdrop—Solutré's perfect twin. The name Vergisson derives from the Celtic word *verg* for "tier" and refers to the tier-like arrangement of the landscape all around this natural amphitheater. The vines occupying the top tiers are logically named Aux Vignes Dessus ("The Vines on Top"), whereas those facing us, on the opposite slope leading up to the Vergisson cliff, are named La Côte ("The Hillslope"), Les Crays ("The Chalks"), and Les Croux ("The Chalky Ones"); the gently sloping back of the rock, facing east, is Sur la Roche ("On the Rock").

Wines from Vergisson are the most mineral, fine, and freshly acidic of the Pouilly-Fuissé family, although the south-facing lots take advantage of the sunnier microclimate to add depth and opulence to the wine's palette. Many critics, like Robert Joseph, find the Vergisson wines, on a par with those from Fuissé, to be the fullest-flavored of the appellation, and those from Pouilly to be the most complex.

THE ROLE OF TOPOGRAPHY

Besides their producers' masterly control of vine-growing and winemaking techniques, the variety and quality of Pouilly-Fuissé wines can be linked to the topography, geology, and microclimate of each lot—three

POUILLY-FUISSÉ AOC

Region:	Mâconnais
Wine type:	dry white
Grape variety:	Chardonnay
Area of vineyard:	750 hectares (1,875 acres)
Production:	44,000 hectoliters/year (5,850,000 bottles/year)
Famous "crus":	many *climats*
Nature of soil:	calcareous, marly
Nature of bedrock:	granite, rhyolite, limestone, marl
Age of bedrock:	lower Jurassic (200 million–175 million years ago)
Aging potential:	3 to 6 years
Serving temperature:	11–13°C (52–55°F)
To be served with:	fish, shellfish, white meat, cheese

factors that are tightly interwoven. The skyward thrusting of the limestone strata built the Mâconnais hill front, which in turn channeled the winds, clouds, and precipitation, and also exposed deep-seated layers that are now the substrate of the Chardonnay vines.

The sedimentary layers were formed in the Early Jurassic, a period beginning 200 million years ago. At the time, France was just emerging from the stable Triassic period, when all the continents were assembled in the giant Pangaea landmass and surrounded by a global ocean. The landmass was shaped like a horseshoe, with its branches pointing east and framing a wide gulf that would later shrink to form the Mediterranean basin. France and the rest of what would become Europe were located at the back of the bay, and although they were exposed to typhoons blowing off the water, their climate had long been continental and dry. As we saw in Alsace, the Triassic period was a time of arid, sandy landscapes and briny coastlines, home to an emerging line of great reptiles.

We find the fossilized tracks of these early reptiles in the Mâconnais region, notably in the township of Verzé, a dozen kilometers (7 miles) north of Vergisson and Solutré. Verzé is known for its Romanesque church and its own brand of Chardonnay white, Mâcon-Verzé; it is one of the twenty-seven regional appellations of a quality considered good enough to feature the name of their producing township. But to geologists, Verzé is most famous for its sandstone quarries. A nature trail takes the visitor to an exposure of stone slabs bearing the tracks of archosaurs, ancestors of both crocodiles and dinosaurs. These reptiles were by this time already 5 meters (15 feet) long and weighed half a ton (1,000 pounds), based on the spacing and depth of the footprints. The direction of the tracks suggests that the reptiles were marching southwest to northeast, which was probably the strike of the coastline in those Triassic days.

At the close of this continental era, the great Jurassic period began with a phase of tectonic rifting that carved through the supercontinent, separating Africa and Europe from North America and giving birth to the Atlantic Ocean. Rift branches close to France served as channels that brought seawater into the basins, including Burgundy.

The climate turned oceanic, warm, and tropical—especially since France at the time was much farther south, straddling 30°N, the latitude of Morocco today. As the sea level rose in Europe, the early lines of dinosaurs branched off the ancestral archosaurs and found refuge on the shrinking tropical islands, while marine reptiles diversified and explored the shallow seas in between. Plankton bloomed in the warm waters and showered the sea bottom with calcareous tests, building up a limy layer spiked with shells and other marine animal remains. Every time the water became deeper and coastal life-forms less abundant, the flaky mineral particles discharged by the rivers dominated the sedimentary process and built up clay-rich layers instead of limestone, or a mix of both called marl. Occasionally, the sediments entombed the large, coiled shell of a deepwater ammonite, or the torpedo-shaped shell of a belemnite, ancestors of our present-day squid and octopi. Later on, when the sea level dropped, coral reefs and oyster banks built up shallow mounds to remain near the surface and live off the sunlight and plankton.

These vast cemeteries of bygone animals are what make up the Mâconnais substrate today, which in places reaches a thickness of over 500 meters (approx. 1,650 feet). But this lime and clay "layer cake" would have lain forever buried, had it not been uncovered by a regional uplift and the ensuing erosion that wore down the land to reveal the ancient strata.

This uplift, responsible for not only the uncovering of the strata but the present-day topography as well, is the result of a tectonic collision between Italy and France that began about 50 million years ago. The collision raised the Alps along the plate boundary, and by domino effect reawakened old fault systems behind the front line: France's ancient Massif Central was also shoved upward, along with its thick blanket of

Triassic and Jurassic sediment. Meanwhile, the in-between crustal section between the Alps and the Massif Central inversely stretched and collapsed, forming a down-faulted basin or graben that today hosts the lake-sprinkled Bresse region, famous for its poultry farming and crossed by the Saône River.

On the Massif Central side, the uplifted Mâconnais range was also affected by the extension. It broke along the splintering fault zone and foundered into a set of steps. Each of these blocks tilted as it slid downward, turning its leading edge skyward to form a ridge, with a cliff dropping down to the next step.

The final touch brought to the landscape was a series of minor faults, perpendicular to the ridge, which carved the hill front into segments with the help of small rivers digging their way down to the Saône valley. The main ridge line, falling abruptly to the west and sloping off progressively to the east, is therefore sculpted into a discontinuous series of hills that also carry north- and south-facing slopes, opening up the Pouilly-Fuissé vineyard's range of sun exposures.

The regional topography also affects the climate—a favorable climate to begin with. The Mâconnais is well positioned: it is the gateway to the southern half of France know as Le Midi, and Mediterranean winds and air masses flow up the Rhône and the Saône valleys as far north as Mâcon. The region is also affected by a strong continental influence, encroaching from the east, that brings cold winters and hot summers. Lastly, humid air

---

SAINT-VÉRAN AND OTHER MÂCONNAIS WHITES

The Mâconnais region is famous for its starring Pouilly-Fuissé wine, but it also hosts four other white appellations that are excellent as well, and more affordable: Saint-Véran, which frames the Pouilly-Fuissé area both north and south; the smaller appellations of Pouilly-Loché and Pouilly-Vinzelles to the southeast; and Viré-Clessé, which stretches south down the Saône valley, almost as far as Tournus.

Saint-Véran, Pouilly-Loché, and Pouilly-Vinzelles wines are not that different from their noble neighbor Pouilly-Fuissé, whereas Viré-Clessé is grown on a southerly segment of the Mâconnais hills. Its marly Jurassic strata encompass a greater age range than at Pouilly-Fuissé, from the very first marls of the Lower Jurassic up to the Oxfordian limestone of the Upper Jurassic, which we will also find further north in the Côte-d'Or hills of Burgundy. Viré-Clessé has the full-bodied character that comes from marl, a bouquet of lime tree and hazelnut, and heralds the great white wines of the Côte de Beaune.

from the Atlantic blows in from the west, but is stopped by the highest ridges of the Mâconnais chain, which intercept most of the rainfall and let only a sprinkle reach the lee side, thereby supplying the desired minimal amount of water to the vines.

VISITING THE JURASSIC SEA BOTTOM

Even the best climate and the most favorable slopes do not necessarily yield great wines. Rock and soil must also contribute the right mineral content, acidity and alkaline balance, permeability, ground texture, and percentage of clay, sand, gravel, and larger chunks of rock. Add to that the almost mystical dimension of the terroir: in the Pouilly-Fuissé vineyard, we tread an ancient sea bottom, a cemetery of limestone debris that was once the tests, bones, and shells of long-vanished sea life. When we hike up a slope, we view this pile of sediments in cross section, from the oldest to the youngest that accumulated on the site—a march through time across millions of years. Each lot we visit and each wine we taste is a new window into France's Jurassic Park.

If we wish to study the Pouilly-Fuissé terroir in chronological order, we must begin our trek at the bottom of the deep hollow mentioned earlier, where erosion has carved the pile down to the basement floor of the Jurassic period. We can start at La Gorge aux Loups ("Wolves' Valley"), off the road that dips down from Solutré toward Vergisson. As you'll recall, the road crosses a stretch of woods; but before reaching the bottom of the hollow we turn right, into a narrow road leading to the hamlet of Chanseron. Completely removed from the already minimal road traffic, its old stone houses and narrow alleys are charged with atmosphere, especially in the fall, when the grapes are pressed and the strong scent of fermenting wine escapes from cellars and open courtyards. The Chanseron hamlet is certainly ancient, as it is built around a Celtic *menhir* ("standing stone"). A paved pathway, polished by centuries of passage, links Solutré to Vergisson across pastures and vineyards: probably a Roman road dating to the first centuries AD. It is crossed by a more recent road for motor vehicles that leads out of Chanseron and is flanked by vineyards: upslope, the lot climbing toward the rock of Solutré is named En Bulland; downslope, Les Verchères stretches down toward the bottom of the hollow.

Along the road is an embankment where the bedrock outcrops (most

Figure 4.3 At the foot of Solutré rock, the Jurassic marl contains many fossils of ammonites (*center*) and ear-shaped oysters (*top* and *right*).

elsewhere, it is buried under soil and vegetation). This rock is a brownish limestone or marl with streaks of gray, but what is most remarkable and noticeable is the amount of fossils that can be spotted with the naked eye: mostly seashells shaped like mussels (fig. 4.3). They are actually oysters from the *Gryphaea* genus, although they are more commonly known as devil's toenails, or *griffes du diable* in French, on account of their curved, clawlike shape. Once your eye grows accustomed to the pattern, you can pick them out everywhere: encrusted in the road embankment, on display in the stone walls lining the vine lots, and broken loose, lying amid the cobbles at the foot of the vines.

Gryphae are typical of the dawn of the Jurassic period, 200 million years ago. Imagine a shallow sea at the time, sparkling under tropical skies, its muddy bottom covered with oyster beds. Add an army of crustaceans and starfish crawling among the oysters, and a few primitive reef fish, swimming to and fro: this sea-life menagerie is what provided the calcareous terroir that Chardonnay vines tap under a few centimeters of topsoil.

If we climb up the slope a few meters, we reach the next level and episode of the Jurassic period. Here we find fewer shells in the sediment and much more clay, an indication that the sea has become deeper and calmer.

This clayey limestone or marl corresponds to the Toarcian stage of the Lower Jurassic, dated around 180 million years ago.

The deeper water was populated by open-sea, larger animals. If you are lucky, you can find in this marl elegant, coiled ammonite shells—ancestors of our squid and octopi. They appear as hollow molds and imprints in rock fragments that rolled down the hillside, and as full fossils, broken out of their coating and lying free in the middle of the vineyards.

Scree deposits are another example of the complexity of terroir. It is not only the local bedrock and its alteration products that make up a soil, but also the rockfalls and alluvium rolling in or washing in from above. Here at La Grange aux Loups, the vines on the slope are fed minerals from both limestone and marl above, a mix that gives the wine a great mineral balance. This holds particularly true for the Chardonnay grape of Pouilly-Fuissé, a variety that does well on limestone but even better on marl, because of the added clay.

In the Pouilly-Fuissé area—and this is also the case in many other vineyards in France—there are two ways to achieve excellence in a wine. For one, the wine coming from a single, excellent lot could be bottled; the lot name would then appear on the label, in smaller characters under the main appellation Pouilly-Fuissé. An example is Pouilly-Fuissé En Bulland, the fossil-rich lot that we just visited.

Another solution would be to blend grapes harvested from two or three different places, if the winemaker is fortunate enough to own a number of lots across a township or even several townships. Such a mix (called *assemblage* by the French) is good insurance against the whims of climate. Lots that present different sun exposures and microclimates will be affected differently by the weather, so the winemakers are able to concoct a satisfactory compromise each year by blending wines off the press that complement each other.

That said, if you want to be practical, how do you select your wine from among 250 producers of Pouilly-Fuissé, who rule over 750 different lots? The best way is to judge the wines yourself at wine-tasting events or by writing down for future reference the exact name of an impressive bottle you tasted at a friend's home or at a restaurant. Better still, visit the villages of Pouilly-Fuissé; the many producers who sell their wine directly to the public more than likely will want you to taste it first.

VINES ON A CORAL REEF

If we resume our geological climb uphill through the Pouilly-Fuissé terroir, we also find in the Toarcian marl fossils of belemnites—another family of mollusks. They were closely related to the ammonites, but lived in
a straight, torpedo-like shell (belemnites owe their name to the Greek
word *belemnos*, which means shaped like a javelin). To catch their prey,
these ancient cephalopods were armed differently from their present-day
cousins: belemnites had hooks on their arms instead of suckers. They measured from tens of centimeters up to 2 meters (about 6½ feet) in length,
but one rarely finds in the Jurassic sediment more than the tip of their inner bone—the most resistant part of the animal's skeleton. These are the
bullet-shaped fossils that vine growers and hunters pick out of the soil in
the marly levels under Solutré and Vergisson.

Alphonse de Lamartine (1790–1869), the leading poet of the Romantic movement in France and a native of the Mâconnais region, compared
the double-pronged escarpment as "petrified boats, shipwrecked in a sea
of vines." This is truly the picture you get when you stand facing the limestone cliffs, their lower walls evoking the prow of some great steamboat,
and their upper part either the boat's smokestack or the turret on top of a
giant submarine (fig. 4.4).

The Solutré rock is a submarine indeed, since it consists of a subaque-
ous pileup of sea life in the Jurassic that never quite reached the surface.
Another 5 million or 6 million years have gone by since the time of the
belemnites, ammonites, and oysters, and the sea level has dropped a few
meters by 175 million years ago as we roll into the Aalenian and Bajocian ages. The site is now under very shallow water—perhaps no deeper
than 10 or 15 meters (approx. 30 to 50 feet)—and the seashore-like fauna
consists mostly of tiny shell fragments, including those of crinoids, also
known as sea lilies. These animals were fixed to the sea bottom with a
stemlike tube from which emerged a bouquet of filtering arms, giving the
animal the appearance of a flower.

When you drink a Pouilly-Fuissé from the upper reaches of the vineyard, namely from the west-facing back of the escarpments, imagine fields
of sea lilies waving back and forth in the current, a current that happens
to be engraved in the sediment as wavy lines that you can pick out in the
cliff limestone.

Figure 4.4 The cliff of Solutré rock displays strata of limestone and marl, capped by a resistant layer of reef limestone—all laid down at the bottom of the Jurassic sea over a time interval of 10 million years. *BP* signifies before present.

After this Aalenian age, the sea over the Mâconnais became deeper and calmer once again. You can notice the trend halfway up the cliff face, where a terrace breaks the slope before the "chimney" resumes a vertical profile up to the top of the rock. Viewed from this terrace, which you can access by a hiking trail branching off the main path, the limestone exposure at the foot of the chimney shows more fragile fossils, such as sponge spicules and even a few ammonite remains.

The uppermost part of the rock, encompassing the last dozen meters of vertical rock face and its erosion-resistant lid, marks another episode of France's fluctuating inner sea: a new lowering of the sea level that occurred around 170 million years ago, during the Bajocian age. This time the conditions were right for the shell bank to host not only sea lilies but also a coral colony, and the mortar of crinoids and crushed coral gave rise to a hard reef mound that now makes up the top of the Solutré and Vergisson rocks. In the vine lots at the foot of the cliffs, you can find honeycomb-shaped bits of coral, shed from above.

On the mounds where it grew, the coral became a hard cap as it aged and fossilized, protecting the underlying rocks from erosion and creating the towering knobs that dominate the landscape today.

With their roots penetrating clay-rich marls, calcium-rich limestone, shattered sea lilies, and coral bits, the vines of Pouilly-Fuissé draw the essence, or at least the memory, of past Jurassic seas. This long and rich geological period witnessed the remarkable evolution of reptiles and especially of dinosaurs that lived on the French islands, which were separated by shallow straits. The sea level's fluctuations repeatedly separated the islands and bridged them back together, spiking the rate of evolution. Now, to pick up the story into the Late Jurassic, we must take our car and drive half an hour north, to the Burgundy hills of Côte-de-Beaune.

CHAPTER FIVE

CORTON AND OTHER WINES OF BURGUNDY

"I serve to my guests a decent Beaujolais wine, but I drink in secret your good wine from Corton," Voltaire confessed to his wine supplier. The witty philosopher (1694–1778), whose books and pamphlets inspired both the American and the French Revolutions, drank *a lot* of Corton. By sheer pleasure or to stay in good health, since wine was tantamount to medicine in those days, Voltaire consumed the equivalent of four bottles a day. We could imagine a worse fate, considering that Corton ranks today in the top list of Burgundy Grands Crus, on a par with Pommard, Clos Vougeot, and Gevrey-Chambertin. Incidentally, Voltaire lived to be eighty-four.

Corton was also a favorite wine of American president John F. Kennedy (1917–1963). It is rumored that Charlie Chaplin (1889–1977) attempted to buy some vine property there, but to no avail. As in other Grands Crus of Burgundy, the terroir is much coveted, so owning vines on a hill such as Corton is a heritage not easily passed on to others.

Corton red wine is served during grand occasions and in top restaurants, where its wild and strong aroma is a perfect match for game and venison—a roasted pheasant, for instance—while Corton white (bearing the distinct appellation Corton-Charlemagne or simply Charlemagne) is considered one of the best white wines in the world. It exhales a bouquet of baked apple, pineapple and citrus fruit, fern and flint. As it ages, a good vintage of Charlemagne also brings out notes of truffle and leather, and develops an explosively rich palate.

As with all great Burgundy wines, Corton reds are made from Pinot

73

Noir, and Corton whites from Chardonnay. Speaking of the latter, the Burgundy Winemakers Association claims that "rarely a grape variety establishes such a tight and graceful link with the character of its terroir."*

Corton is above all a magical place, a hill so well structured that it offers excellent soil, bedrock, slope and drainage, sun exposure and microclimate. Legend has it that it was Emperor Charlemagne (circa 747–814) who came to notice that snow melted more rapidly on the hill than elsewhere—the proof of ideal sun exposure—and ordered that vine be planted there. According to other sources, his grandfather Charles Martel (686–741), a founding figure of the Middle Ages, already owned vines on the hill, which his grandson inherited.

Be that as it may, Charlemagne, who did much during his reign to support monasteries, in 775 donated (at the young age of twenty-eight) his few acres of Corton vines to the monastery of Saint-Andoche in Saulieu,† turning over to the religious order the tending of the precious terroir. Incidentally, besides showing devoutness and generosity, supporting the monasteries helped Charlemagne spread wealth, culture, and control over his rapidly expanding territories.

As was the case elsewhere in France, monasteries applied scientific rigor to the art of winemaking, crossing and selecting the best varieties of grapes for their terroir. In Burgundy, they focused on Pinot Noir, and crossed it with an ancient variety of grape named Gouais to obtain Chardonnay. They also perfected grafting and pruning, and improved winemaking techniques. Moreover, they understood the importance of location, processing grapes from different sites in separate vats, hence developing the concept of "climate." In Burgundy, climate defines a precise growing place, characterized by its soil and climatic environment—essentially a subset of a terroir.

There are twenty-five such climates on the Corton hill, covering no more than 100 hectares (250 acres)—Central Park in New York is three times the size. The name of the individual climate will often be featured

---

* See http://www.vins-bourgogne.fr/.
† The original Carolingian church was destroyed by wars and pillaging during the ninth and tenth centuries. It was replaced by a Romanesque church in the twelfth century that was restored in the eighteenth. Capital of the region of Morvan, Saulieu and its basilica are an hour's drive from Beaune and Corton.

on the label of a Corton, since it usually confers to the wine a very distinct personality. Corton Les Renardes is said to be rough and gamy; Corton Clos du Roi, flanking it to the west, is powerful and velvety; Corton Les Bressandes, a few meters downslope, is full and smooth; and Corton Les Perrières, to the west of Bressandes, is delicate and feminine.

In practice, you might not wish to host a Corton wine-tasting party on your own, since a single bottle of a young vintage costs about 40 euros on-site, 60 euros at your local wine store, and well over 100 euros (or more than 130 US dollars at time of writing) at a fancy restaurant. But if you happen to travel through Burgundy, go visit one of the three villages on the hill (Pernand-Vergelesses, Aloxe-Corton, or Ladoix), follow the signs to one of the many cellars in town, and get treated to a round of wine tasting. Burgundian winemakers have a true sense of hospitality, and they are proud to show off their wines as well as their knowledge of the terroir.

Besides the Corton Grands Crus, the estates usually produce more affordable wines as well, from the lower slopes and the adjacent hills, where they also own vine lots. You can purchase an excellent Aloxe-Corton red, Savigny-lès-Beaune red, or Pernand-Vergelesses white for less than 20 euros, and a good Bourgogne red or Bourgogne Aligoté white for less than 10 euros.

ALOXE-CORTON AND CORTON AOC

Region:	Burgundy, Côte d'Or
Wine type:	rich reds, opulent whites
Grape variety:	Pinot Noir and Chardonnay
Area of vineyard:	255 hectares (637.5 acres)
Production:	10,700 hectoliters/year (1,423,000 bottles/year)
Grand Crus:	Corton
	Corton-Charlemagne, Charlemagne
Nature of soil:	calcareous and pebbly
Nature of bedrock:	marl and limestone
Age of bedrock:	Jurassic (Bathonian/Oxfordian, 165 million–155 million years ago)
Aging of wine:	5 to 20 years
Serving temperature:	14–16°C (57–61°F)
To be served with:	game, venison, beef, cheese (reds)
	lobster, fine fish (whites)

Visiting the Corton hill and its surroundings[‡] is therefore a true delight for amateurs of Burgundy wines and also for geologists, since it illustrates perfectly the importance and intricacies of terroir. In addition, its sediments pick up our geological history of France in the Middle Jurassic period, right where we left off in the previous chapter, in the hills of Pouilly-Fuissé.

THE HILL AND THE WINEMAKER

The Corton hill is called a mountain (*montagne de Corton*) because it is so famous, but it is merely one of the many hills adorning the western shoulder of the Saône valley. The hill front, which trends roughly north to south, is known as the Côte d'Or. This name doesn't mean "Coast of Gold," as you'd expect, but rather "Coast of Orient": the hill faces the rising sun. Its lineup of vineyards is interrupted midway by a notch in the hill front that was gouged out by limestone quarries, dividing the Côte d'Or into two halves: Côte de Beaune to the south (fig. 5.1) and Côte de Nuits to the north.

The hill front was created by a great fault system that slices through the Earth's crust from Dijon down to Le Creusot. The block east of the fault zone was down-dropped to form the Saône valley—also called the Bresse graben—whereas the western block was uplifted to form the Côte d'Or hills, extended to the south by the Mâconnais hills that we covered in the previous chapter.

On account of the Bresse graben, the Saône River flowed southward along the fault lines. Rainfall in the hills, dumped by the westerly winds blowing in from the Atlantic, established a system of confluents flowing perpendicularly to the main valley. Hence the Côte d'Or is divided into segments, separated by rivers and brooks. Five kilometers (3 miles) north of Beaune, the hill of Corton was sculpted by such brooks—now vanished—that separated it from the adjacent hills of Savigny-lès-Beaune and Pernand-Vergelesses. All three hills carry quality vineyards, but it is Corton that has by far the best layout, with a remarkable range of exposures all around its circumference: east, south, west, and even northwest.

[‡] The hill of Corton is a ten-minute drive from the city of Beaune—itself a two-hour train ride or a three-hour drive from Paris. Beaune is one of the wine centers of Burgundy, and its medieval hospital and chapel (Les Hospices de Beaune) are well worth visiting.

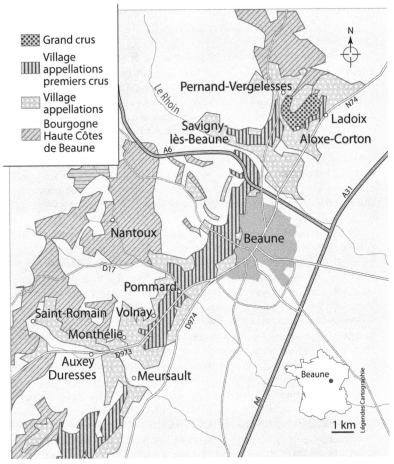

Grand crus

Village appellations premiers crus

Village appellations

Bourgogne Haute Côtes de Beaune

Pernand-Vergelesses

Ladoix

Savigny-lès-Beaune

Aloxe-Corton

Nantoux

Beaune

Pommard

Saint-Romain Volnay

Monthélie

Auxey Duresses

Meursault

Beaune

1 km

Figure 5.1 Burgundy's Côte de Beaune, from Ladoix to Meursault. The D 974 road approximately follows the fault that separates Jurassic strata to the west from more recent basin-filling alluvium to the east. (Map by Legendes Cartographie; modified from the map of the BIVP agency—Bureau interprofessionnel des vins de Bourgogne.)

Only the northeastern section, in the shadow of the summit, is unfit for vine growing.

From the bottom upward, the hill is shaped like a hat, with a broad brim blending into the flat valley floor and followed by the D 974 main road. To the east of the road, the vineyards lie on flat land and yield ordinary Burgundy (Bourgogne AOC [Appellation of Controlled Origin]). To the west, facing the hill, the gently rising slope carries the vines of Aloxe-Corton reds. As the slope steepens, uphill from the village of Aloxe-Corton, the wines are entitled to the appellation Aloxe-Corton

CÔTE DE NUITS

The Burgundy Côte d'Or is divided into Côte de Beaune, south of the medieval town of Beaune, and Côte de Nuits to the north, stretching as far as Dijon. Côte de Nuits owes its name to the central village of Nuits-Saint-Georges, nestled in a valley that indents the hill front. All along this northern half of the Côte d'Or, many villages are home to famous Grands Crus: Gevrey-Chambertin, Morey-Saint-Denis, Chambolle-Musigny, Vougeot, and Vosne-Romanée.

Along the great sequence of undulating limestone and marl strata that make up the Burgundy terroir, the Côte de Nuits corresponds to an anticline, centered on Gevrey-Chambertin. It is therefore the uplifted lower part of the sedimentary sequence that outcrops in the vineyards, namely the limestone and oyster marls of the Bajocian age (Lower Jurassic, 172 million to 168 million years ago), topped by the handsome pink limestone of Prémeaux. Above that comes white limestone, capped by the hard, resistant Comblanchien limestone that forms a steep ledge overlooking the vineyards.

Because it is capped by hard limestone, the Côte de Nuits has a steeper slope than the Côte de Beaune to the south, because in the latter the hard rock—susceptible to breaking in ledge-like fashion—is buried below the surface. Steeper slopes in the Côte de Nuits mean a narrower strip on which to grow vines. In places, however, the narrow hill front is broken up by perpendicular valleys; these dump broad cones of alluvium outward into the Saône valley, which extends the vine-growing area. This is the case, for example, at the foot of Gevrey-Chambertin.

Gevrey-Chambertin, grown at the northern end of the Côte de Nuits, was Napoleon's favorite wine (he took it with him even to the battlefield). Of the forty Grands Crus of Burgundy, nine belong to the appellation Gevrey-Chambertin (such as Chambertin-Clos-de-Bèze and Griottes-Chambertin). They are complex, full-bodied wines with aromas of plums and cherries, and should age at least five years in the cellar.

Chambolle-Musigny, 5 kilometers (3 miles) to the south, has the reputation of being the most "feminine," fine, and delicate Burgundy red of the Côte de Nuits, in the same way Volnay is the most feminine wine of the Côte de Beaune. The finesse of the wine is credited to the limestone strata jutting out into the vineyards, and the fact that they are highly fractured, enabling deep root penetration and perfect hydraulic and mineral balance. One of the two Grands Crus of Chambolle-Musigny is Bonnes Mares, north of the village, which taps white, clay-rich marl, giving the wine exceptional opulence and body; it needs to age over five years to deliver its full gamut of aromas. On a terrace south of the village, Musigny is the other Grand Cru. It manages to strike a balance between robustness and subtlety, giving off a bouquet of red berries, white flowers, and oriental spices.

Vougeot and its famous Grand Cru Clos-de-Vougeot lie directly below the terrace of Musigny. Here the rock is a lower and hence older Bajocian limestone (sea-lily limestone). The slope is less steep, with the soil becoming thicker and richer in clay. Vougeot provides a good example of terroir complementarity within a single vineyard. Here, the estates often blend grapes from the upslope lots, which have the complexity and finesse provided by limestone, with grapes of the lower slopes, where clay provides body and richness to the wine. The end result is a mineral, yet sophisticated wine with great body and aging potential (five to twenty years), developing over time a bouquet of prune and black cherry, violet and truffles.

Vosne-Romanée possesses eight Grands Crus within an area half the size of New York's Central Park, including the famous La Romanée and Romanée-Conti. These two might owe their names to Roman ruins discovered on the site; as for the suffix *Conti*, it refers to Prince Louis-François de Bourbon-Conti (1717–1776), diplomatic councilman of King Louis XV, who bought the vineyard in 1760. Elegant and velvety, with a rich bouquet of prune, all Vosne-Romanée Grands Crus are raised on the lowest strata of the Côte de Nuits sediments: *pierre de Prémeaux*, a pink limestone brought up to the surface at this level by the regional anticline, overlain by white granular limestone of Bathonian age, with some marl mixed in. Minor faults break up the terroir further, thereby multiplying the number of special climates and personalities of Vosne-Romanée wine: a bouquet of wild roses at Romanée-Conti; black cherry at La Romanée; violet at Romanée-Saint-Vivant; licorice at La Tâche; raspberry at Richebourg; and black currant, pepper, and forest undergrowth at Échezeaux.

Figure 5.2 Harvest halfway up the slope of the Corton hill, on the "climate" known as Les Renardes. In the background, the hill to the right separates the vineyards of Pernand-Vergelesses (*near side*) and Savigny-lès-Beaune (*out of the shot, far side*).

Premier Cru—a jump in quality—and halfway up the hill begins the terroir of Corton Grand Cru (fig. 5.2), with Pinot Noir lots facing east to south, and Chardonnay curling around the western face of the hill. The bracket of elevations ranges from 250 to 350 meters (approx. 830 to 1,160 feet) above sea level. At the very top, the climate is cooler and the limestone blocky; unfit for cultivation, the hillcrest is capped by forest.

You'll catch this global picture if you drive toward the hill on route D 974, heading northward out of Beaune toward Dijon. Past the low stone walls that bound the vineyard, you'll notice a deep-red color to the soil, caused by a layer of iron-rich limestone that spread its red oxides down into the plain.

As you near the hill, a side road branches off to the west, heading for Pernand-Vergelesses. This quaint little village wears two hats: it offers its own appellation, Pernand-Vergelesses AOC, which includes some distinguished Premier Cru white Chardonnays (Sous-Frétille, Clos Berthet); what's more, its district includes the western flank of the Corton hill and boasts many lots of Corton-Charlemagne Grand Cru.

Yet you would never know any of this from Pernand-Vergelesses's modest appearance and quiet atmosphere. There, as in most Burgundy villages, wealth is not flaunted, and no châteaux are in evidence—not even a coffee shop or a grocery store, for that matter—only a little Romanesque church and a few elegant rooftops of polychrome-glazed tiles. If it weren't for the signposts, you might not even suspect that great wine cellars were here.

Just below the church, the house of Eric Marey has a flight of steps leading down to a small cellar containing a long table, cases of wine everywhere, and a painting of Eric's father, Pierre Marey, standing guard over a pile of stacked bottles. Here, Eric will have you sample each wine from his production, which runs the gamut from inexpensive Bourgogne and Bourgogne Aligoté, through medium-priced Pernand-Vergelesses red and white, up to Corton and Corton-Charlemagne Grands Crus. You need not buy a single bottle, but I can't imagine anyone walking out of this cellar empty-handed.

As with other great terroirs, there are two ways to make a great wine in Corton. One is to vinify grapes from only a single renowned "climate" (the name of which will be specified on the bottle label) that brings out the essence of its particular terroir. The other method is to blend grapes from several climates to achieve a good balance. Eric Marey's Corton Grand Cru is produced by blending—hence no subtitle on the label— but should you ask, you will learn that it combines two of the best climates on the hill: Clos du Roi and Les Bressandes.

If you wish to compare climates, walk across the village to the cellar of the Dubreuil-Fontaine estate. Here you will get a complete rundown of the many lots the family owns on the hill and elsewhere (including Pommard)—glass in hand, of course—and you will be able to taste and purchase separately Corton-Clos du Roi and Corton-Bressandes, although their subtle differences need a few more years in your cellar to truly blossom.

Should you wish to learn more about making a great wine, take a walk to the Denis estate at the top of the village. The presses and winemaking vats are on-site, and Christophe Denis will be happy to take you on a tour of the production chain before opening up a few bottles. His range includes red Savigny-lès-Beaune from around the hill, white Pernand-Vergelesses, and, of course, Corton and Corton-Charlemagne.

The making of a red Burgundy begins at harvesttime—from late August to mid-September, depending on the weather—when bunches of Pinot Noir grapes are packed in cases, then brought to the winery and discharged onto a conveyor belt. Next, the grapes travel past a sorting table, where the staff, including Christophe and his father, removes any sad-looking ones or overlooked leaves or stems. The grape bunches then drop into a rotating drum, where paddles dislodge the individual grapes from the stems. Now the grapes fall through slots in the drum, onto another conveyor belt that climbs upward before toppling them into the maceration vat—a metallic container in which fermentation takes place. Fermentation is the job of bacteria naturally present in the grape juice, which can also be spiked with concentrated bacterial culture from previous fermentation runs.

The bacteria, also called yeast, transform the grape sugar into alcohol in only a few days. This process is under the close scrutiny of the Master Blender, who monitors the juice's rising temperature—a consequence of the bacteria's frenetic activity. Should the temperature rise over 40°C (100°F) or so, cold water is sent through coiled tubing to circulate around the vat, removing excess heat and stabilizing the juice temperature (the same technique used in cooling rocket engines).

In the case of red wine, the maceration of grapes continues in the vat for approximately two weeks so that tannin, other aromatic molecules, and color transfer from the grape skins into the juice. The young wine is then drawn from the vat and transferred to oak barrels, where it remains for about a year, gathering more tannin from the oak wood itself before being bottled.

The production of white wine is a different process. The juice must not spend too much time in contact with the skins, to avoid picking up color and tannin. The grapes are therefore pressed as soon as they come off the conveyor belt, in a rotating drum that contains a central, inflatable bladder. The bladder crushes the grapes against the wall of the drum and the juice is expelled through its slits. If the grape variety is Aligoté, the juice is transferred to a metallic vat in order to preserve its light, fruity character. If the grape is the nobler Chardonnay, the juice is transferred to an oak barrel in order to absorb some tannin from it, which will help the aging process and instill extra woody aromas, such as vanilla.

THE NOTION OF "CLIMATE"

In a winery, each prized "climate," or "Cru," of the estate is entitled to its own vat, of a size proportional to its acreage. In the Denis estate, there is a large vat for their Savigny-lès-Beaune red wine—several hectares' worth of vines—whereas the Pernand-Vergelesses harvest fits in a medium-sized vat, and the Corton grapes in the smallest container, reflecting that lot's limited and precious acreage.

Approximately six hundred numbered lots occupy the Corton hill, each averaging one-fifth of a hectare (half an acre) and split among approximately seventy winemakers. Often, a winemaker will own lots in different "climates" on the hill, as do Eric Marey and Dubreuil-Fontaine, who own at both Clos du Roi and Les Bressandes. The twenty-five climates at Corton, encompassing a few dozen lots apiece, were defined over the years as entities having distinct soil, slope, altitude, and sun exposure, with the latter two characteristics justifying their very name of climate.

As we saw earlier, the vineyards wrap around the eastern, southern, western, and even northwestern sections of the Corton hill. You might think that the climates facing due south or southeast—such as Clos du Roi and Les Bressandes—get more sun and hold an advantage, and they do if you are looking for the most powerful reds. But climates with a western exposure also have their good points, including slower, more aromatic maturing of the grapes, which is a plus for Chardonnay. For this reason, the climates assigned to the making of white Corton-Charlemagne occupy the western section of the hill, as well as the highest, coolest elevations.

Besides exposure, slope is key in draining the soil if rainfall is excessive. It is a fact that for the vines to produce both plentiful and quality grapes, the plant needs to be water stressed: it has to be kept thirsty. To promote runoff, the vines are planted in lines running downslope rather than cross slope, so that the rows channel the water downhill instead of barring its way.

This planting geometry might look strange to a farmer, since encouraging runoff of water also causes precious soil to be carried away—in agriculture, it is customary to plant cross slope to combat this erosion. Clearly, winegrowers have made a choice and favor water evacuation. In particular, the morning fog, which forms in the wooded area capping the summit, can roll down the hill unimpeded, rather than hang over the rows and threaten to rot the grapes.

The price to be paid is soil depletion, and the winegrowers compensate by periodically hauling soil and stones up the hill. A natural but much slower contribution is made by the top limestone layer at the edge of the woods; it slowly erodes and sheds rubble onto the slope.

A TROPICAL LAGOON

The precious rocks of Côte d'Or are a combination of limestone and marl (limestone spiked with clay), sediments that were deposited in the shallow sea covering up Burgundy during the Jurassic period, 200 million to 150 million years ago (fig. 5.3). Islands and beaches, lagoons and coral reefs were sprinkled across the tropical waters, so that the nature of the sediment varies from place to place, as it also does across time from one geological age to the next. This explains the variety in soil and bedrock both horizontally and vertically, and why "climates" are so diverse on such a small scale.

You can get a good view of the bedrock in cross section if you drive up route D 974 a couple of kilometers (1¼ miles) north of Corton, and proceed through the townships of Corgoloin and Comblanchien. Here, the rock strata have been quarried to access and exploit a limestone bench, roughly 10 meters (33 feet) thick, that provides a handsome, marble-like ornamental stone which is beige with streaks of pink. This Corton or Comblanchien stone adorns many historic French buildings, including the façade of the Paris Opera and the paved floors and stairways of the Louvre.

When you stand looking at the rock face in a quarry, you are actually beholding millions of years of sedimentation—the slow accumulation of

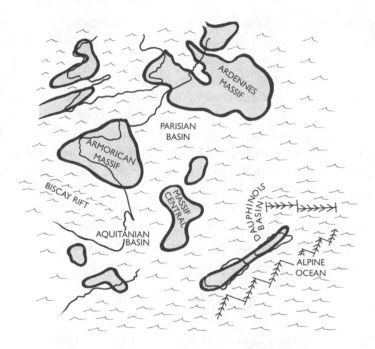

Figure 5.3 France during the Jurassic period, 150 million years ago. Only a few elevations emerge from an inland sea covering most of Europe. Burgundy lies under shallow tropical waters. (Map by Pierre-Emmanuel Paulis.)

scores of small shells and plankton tests that had sunk to the bottom of a prehistoric sea. This is the recipe for limestone. The clay-rich variety known as marl is formed when rivers dump clay particles into the sea, where they are then mixed up with the calcareous material. Depending on water depth and the abundance of plankton and other marine life-forms, marl and limestone accumulate on the seafloor at rates of several millimeters to a few centimeters per thousand years, so that it takes one hundred thousand years on average to build up a rock layer one meter (approx. 3 feet) thick.

In a Comblanchien quarry, the limestone strata are 60 meters (200 feet) thick—roughly 6 million years' worth of sedimentation—but only the central 10 meters (33 feet) have the proper structural quality to be marketable for building and paving purposes. This fine-grained bench is made up of microscopic algae concretions that indicate they were deposited in very calm waters, probably a lagoon. Sea worms burrowed through the soft ooze at the time, expelling trails of filtered particles as they went, so that much later, after millions of years of compaction and mild heating at the bottom of the rock pile, the tiny burrows stand out as pinkish veins that give the limestone its marble-like aspect.

Above the lagoon limestone bench, the rock takes on a coarser texture and a deeper beige color laced with brownish cracks, over a range of 10 meters or so. Its warmer hue also qualifies it as an ornamental stone, used for paving, windowsills, and chimney stones.

At the top of the sequence, the last few meters of quarried material is Ladoix stone, named for the nearby village on the Corton hill, but it is commonly known as *lave de Bourgogne* ("Burgundy lava") because of its hardness and reddish-brown color. Rustic-looking and resistant to weathering, it is used in restoring historic monuments and building outdoor terraces and swimming pool decks. Ladoix stone is also found in most of the walls bordering Burgundy vineyards, including those of the famous Clos-Vougeot.

The entire rock sequence is also referred to as *la dalle Nacrée* ("the pearly slab") because of the shiny fossils embedded in the stone: sea lily stems, oyster shells, and sponge fragments.

Although it is laid bare in the quarries, the *dalle Nacrée* actually underlies the vineyards in the plains and lower slopes of Aloxe-Corton and Pernand-Vergelesses. The vine roots thus tap calcareous sediments of a sea that covered the area 165 million to 160 million years ago, during the Bathonian and Callovian ages of the Jurassic period. Only at the foot of the slopes is this bedrock covered by fairly thick soil, fed by runoff and rockfalls from above and containing clay, limestone pebbles, and even flint. The combination of limestone bedrock and deep, relatively damp, pebble-stuffed soil yields full-bodied Burgundy wines—for instance the Aloxe-Corton AOC on the lower slopes of the Corton hill—characterized by aromas of cherry, blackberry, and black currant, evolving toward prune, truffle, and leather.

CLIMBING JURASSIC HILL

As we leave the *dalle Nacrée* flooring and head up the hill of Corton, penetrating the terroir of the Grands Crus, we notice a change in the

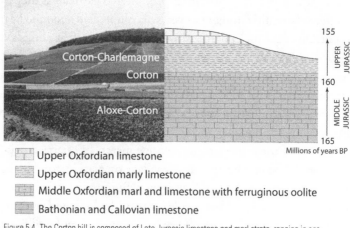

Upper Oxfordian limestone

Upper Oxfordian marly limestone

Middle Oxfordian marl and limestone with ferruginous oolite

Bathonian and Callovian limestone

Figure 5.4 The Corton hill is composed of Late Jurassic limestone and marl strata, ranging in age from 165 million to 155 million years before present. Three distinct wine appellations share the slope: Corton-Charlemagne, Corton, and Aloxe-Corton (*from top to bottom*). *BP* signifies before present. (Map by Legendes Cartographie.)

sedimentation history of the Burgundian sea (fig. 5.4). The last layers of pearly limestone display a wavy form of layering, a sign that strong currents flowed over the sea bottom at that time.

The sequence then ends with a meter-thick layer of reddish oolite, or "egg rock," so named because the sediment is made up of tiny spherical concretions of limestone (and occasionally other minerals) roughly one millimeter in diameter—the size of small fish eggs. The ferruginous layer probably signals strong intertidal currents at the time, with entrained particles receiving coatings of calcium and iron until they became heavy enough to drop to the sea bottom. At Corton, the layer of ferruginous oolite tints the entire hill beneath it with red ochre.

Sediments at this level contain fossil fragments of ammonites that occasionally got entombed in the mineral mud—those cruising cephalopods with spiral shells that we already encountered in the Mâconnais hills of Pouilly-Fuissé. The specimens from Burgundy are later, more ornate species, including *Ammonites cordatus*, with a sinuous shell evoking a coiled rope, and *Ammonites perarmatus*, with bristling knobs that certainly discouraged many a predator.

And fierce predators in the Jurassic seas were particularly fond of ammonites—not the shell part but the squid-tasty occupant. Judging from the teeth marks found on fossilized ammonite shells, we know that the creatures were hunted down by great marine reptiles, especially the long-necked plesiosaurs and the dolphin-shaped ichthyosaurs. Their fossils have not yet been found near Corton proper, but a well-preserved ich-thyosaur skeleton was discovered in the Jurassic marls near Dijon, a little farther north. Imagine as well giant marine crocodiles roaming the waters and flying reptiles gliding through the air—pterodactyls with a 5-meter (16-foot) wingspan and the feathered ancestors of birds—and you will get a broad picture of Burgundy's Jurassic lagoon.

Above the fossil-rich red clay begins a new sequence of massive lime-stone that records shallower water and a return to coastal-like conditions at Corton. The light gray to slightly pinkish limestone contains a new load of sea lily and shell fragments along with small amounts of clay—a calcareous terroir that continues to favor Pinot Noir and is home to the red Corton Grands Crus.

Finally, as we move up the geological time scale one more notch and reach the top of the hill, the clay content increases to the point where the sedimentation layers are now called marls rather than limestone. Al-though Pinot Noir is not overly sensitive to the difference and doesn't need clay to excel, the white Chardonnay grape does favor the change. Besides the slightly cooler microclimate at this level, clay content is the main reason why the upper lots of the hill are planted with Chardonnay and yield the superb Corton-Charlemagne whites.

The name of the exact climate is not specified on the label of a Corton-Charlemagne, contrary to Corton reds. It's a matter of tradition. But an experienced oenologist will recognize a western-slope Corton-Charlemagne as being more mineral and having more bite than one from the southern slope, which will have a fuller body and need more years than its neighbor to reach its prime.

FROM POMMARD TO MONTRACHET

Considering the minute differences and subtleties of a single appellation across 100 hectares (250 acres) or so of vineyards on a single hill, it is no wonder that the Grands Crus of Burgundy offer such a range of wine-tasting experiences when one takes into account the dozens of hills and villages that dot the Côte d'Or.

From Corton, we need drive less than 15 kilometers (a little over

9 miles)—under 15 minutes—to reach Pommard and Volnay, two of the most famous appellations of the Côtes de Beaune. Separated only by an invisible cadastral boundary, their adjacent vineyards yield markedly different wines: Pommard is known as one of the most robust, masculine Crus of Burgundy, whereas Volnay is considered one of the most feminine—more supple and fragrant.

At first sight, Pommard and Volnay share the same terroir, but their contrasting characters derive from minute differences in slope, exposure, rock strata outcropping at the surface, and scree cover sliding in from above.

Broadly speaking, the bedrock geology is the same pile of marl and limestone found at Corton, but a structural factor also comes into play. Since the time when they were laid down horizontally in the Jurassic sea, the sedimentary layers have been deformed by tectonic forces into a gently undulating waveform. From north to south, the strata first bulge upward, bringing the deeper layers to the surface in the vicinity of Gevrey-Chambertin (geologists call this segment the Gevrey *anticline*). The strata then dip downward, and it is their upper layers that end up flush with the vine-growing altitude near Volnay (the Volnay *syncline*). Finally, they curl upward once more in the southernmost villages of the Côte d'Or, returning deep-seated layers to the surface around Meursault and Montrachet.

Any given layer of the sedimentary series is therefore present at different elevations, depending on the township, and will or will not be part of the vine-growing belt. In the vicinity of Corton, the top of the pearly limestone sequence (the *dalle Nacrée*) outcrops 260 meters (866 feet) above sea level—flush with the lowermost vines of the appellation—whereas at Volnay and Pommard it is 20 meters (66 feet) lower and out of range of the vines.

At Pommard and Volnay, it is the overriding younger strata that are flush with the vineyard—those that at Corton were higher up, level with the forest. As for the youngest strata at the top of the slopes, there are some that aren't even found at Corton: clay-rich marls known as *marnes de Pommard*.

As is the case on the Corton hill, the best vineyards at Pommard and Volnay grow halfway up the hillslope, where exposure and drainage are optimal. As for orientation, the main hill front also faces the rising sun, since the same northeast-to-southwest fault system governs the landscape.

At Pommard, a perpendicular streambed cuts the hill front into two wings. Known as the Grande Combe, the valley hosts the village itself (a population of less than five hundred souls), which nestles around an imposing church. Right below it on the flatlands stands the Château de Pommard, one of the rare châteaux of the Burgundy vineyard. On either side of the valley, different exposures again yield different wines. In the northern wing, which faces south to southeast and has a steep slope, the wines are fine and elegant. The southern wing faces mostly eastward and its wines are more robust, in line with Pommard's rather "manly" reputation.

At its southern border, Pommard merges invisibly across the township line with the adjacent terroir of Volnay. Wines from this Volnay fringe are logically similar to those of Pommard, but after proceeding a few hundred meters, as we near the village of Volnay, a more "feminine" personality takes over, with fragrant aromas of raspberry and violet and a more supple and balanced spirit. The farther south we proceed, the more obviously this fragrance and smoothness take over, which make Volnay one of the most seductive wines of Burgundy.

We have now reached the southern tip of the Côte d'Or and the village of Meursault, known for its spectacular Chardonnay white. While other villages of Burgundy deliver only a small fraction of white wines—though Corton-Charlemagne is an exception, producing some of the very best in the world—Meursault and its neighboring appellation of Puligny-Montrachet are almost exclusively devoted to the Chardonnay grape, and apparently have been for many centuries. Thomas Jefferson (1743–1826), ambassador to France before becoming the third president of the United States and an astute observer of vineyards—in Virginia he started the first winery in America with British entrepreneur Philip Mazzei—noted that "at Meursault only white wines are made, because there is too much stone for the red."[§]

A rocky slope is one thing, but Meursault and Montrachet are essentially home to Chardonnay because the wavy structure of the rock strata, as noted earlier, after bending down near Volnay, dips upward at this level and brings to the surface the lowest levels of the sedimentary pile. These layers happen to be marls rich in clay—a perfect match for the blond grape.

This constant oscillation between pure, metal-poor limestone and richer

[§]Jefferson manuscript, 1787; quoted in Robert Joseph's book *French Wine* (New York: DK Publishing, 2005).

marl at the bottom of the Jurassic sea, and the amount of uplift affecting which level is now flush with the vineyards, is what governs the nature and variety of Burgundy wines. What happens next in the grand story of the Jurassic period is to be found—and tasted—farther west by jumping to the Loire valley and the younger sediments of Sancerre.

CHAPTER SIX

Sancerre and the Upper Loire Valley

As we drive up the Loire valley, half an hour before reaching the city of Nevers, we will cross the Sancerre winegrowing district, on the edge of an old French province known as Le Berry. In this peaceful agricultural district, the history of the vineyards is closely intertwined with that of France.

Perched on a hill, strategically located at the junction of the river valley with major Celtic and Roman roads, Sancerre was destined to become a military stronghold, as it overlooks both the Loire River to the east and cross valleys penetrating the Berry plateau to the west. In the twelfth century, Sancerre's ruler Etienne I built a fortress on the crest of the hill, and a rampart midway down the slope to protect the city dwellings. It so happened that Etienne was also Grand Bouteiller de France, in charge of the royal table and wine cellars, which certainly helped promote Sancerre wine among the French nobility.

As far as the wine's production was concerned, it was handled by Augustine monks from the nearby Abbey of Saint-Satur. The vineyards were well established by then, since Gallo-Roman bishop and historian Gregory of Tours mentions them in his writings as early as AD 582. Other vineyards farther west, up on the Berry plateau, were controlled by a rival monastery of Benedictine monks and eventually gave rise to a different wine appellation. Similar to Sancerre in taste and quality, it is known today as Menetou-Salon.

We can well imagine the boisterous rivalry between monasteries, merchants, and city nobles over the lucrative commerce of wine; but the history of Sancerre soon took a much darker turn with the onset of the French Wars of Religion (1562–98), which pitched Catholics against Protestants. The latter's revolutionary doctrine was taught in France by John Calvin (1509–1564), a humanist lawyer who studied at the University of Bourges—the main city of the Berry region, 50 kilometers (30 miles) southwest of Sancerre.

Calvin started spreading his heretical ideas in the late 1520s. Over the next few decades, so many Huguenots moved to the region to escape persecution elsewhere in France that the city of Sancerre became known as La Petite Rochelle. The nickname refers to the other Protestant stronghold at the time—the fortified harbor of La Rochelle on the Atlantic coast.

Following the killings of French Protestants in Paris in August of 1572 (known as the St. Bartholomew's Day Massacre), persecution and violence spread to the provinces. The Catholic League attempted to take the city of Sancerre by storm on the evening of November 9, but the inhabitants managed to repel the invaders after an all-night battle. It took an eight-month siege by the royal army before the starving Huguenots finally surrendered, in August of 1573. The ramparts were dismantled, Sancerre was plundered, and French king Charles IX even went so far as to seek damages and taxes from the ruined city to pay for the siege, in the form of 2,000 liters (over 500 gallons) of Sancerre wine.

It was another twenty-five years before peace returned to the province of Le Berry, thanks to the religious compromise embraced by the new king, Henry IV. By signing the Edict of Nantes on April 13, 1598, the French monarch authorized freedom of worship. If one reads the fine print, however, a clause specific to Sancerre allowed the local Huguenots to congregate only in smaller, neighboring villages, "and not in the city proper."

The vineyards of Sancerre managed to survive those somber days of French history, mostly because the wine they produced was so highly praised and a treasure to be spared. Legend has it that Henry IV himself, setting up camp at the foot of the hill and sampling the local beverage, proclaimed to his army: "By God, this wine is the best I have ever tasted! If all in the kingdom were to drink it, there would be no more wars of religion!"

SANCERRE AOC

Region:	Sancerrois (Centre Loire)
Wine type:	white, rosé, red
Grape variety:	Sauvignon, Pinot Noir
Area of vineyard:	2,800 hectares (7,000 acres)
Production:	165,000 hectoliters/year (21,950,000 bottles/year)
Crus:	none officially, but famous climates (e.g. Les Monts Damnés)
Nature of soil:	marly *terres blanches*; limestone *caillottes* and *griottes*; clay and sand with flint
Nature of bedrock:	limestone, marl, minor sand
Age of bedrock:	Jurassic (Oxfordian, Kimmeridgian, 160 million–150 million years ago) Cretaceous and Tertiary east of the Sancerre fault
Aging potential:	1 to 4 years
Serving temperature:	10–12°C (50–54°F)
To be served with:	white: fish, chicken, duck, goat cheese red: poultry, rabbit rosé: cold cuts, Asian cooking, cheese

FROM PINOT NOIR TO SAUVIGNON BLANC

In the olden days, Sancerre was a red wine, made from Pinot Noir. This was the rule in the region, ever since Philip the Bold (1342–1404), Duke of Burgundy, had imposed Pinot over Gamay, a grape variety that he judged inferior and "disloyal" (see chapter 2). Pinot Noir did very well in Sancerre, and Philip's neighbor John the Magnificent (1340–1416), Duke of Berry, judged its wine "the best in the kingdom."

Sancerre's political and commercial influence waned somewhat after the religious wars, the sacking of the city, and the exodus of Huguenot merchants and noblemen, but its red wine continued to be held in high esteem. On the eve of the French Revolution, which would ultimately topple him, King Louis XVI regarded Sancerre wine as one of his favorites. His love for it was reciprocated in that the region remained a royalist stronghold throughout the revolution and participated in the Chouannerie uprising against the Republic, which was ultimately subdued.

In surviving so many tragedies and misfortunes, the Sancerre vineyard appeared to be indestructible. It was further bolstered by new trade routes that opened up: the Lateral Canal along the Loire River in 1838, and the Bourges-Sancerre railroad in 1885. But an ultimate blow was to hit the vineyard, in the form of a foreign, tiny pest—the phylloxera moth and its larva—that reached Sancerre in 1886, twenty years after crawling or flying off boats from North America that had delivered produce to the harbors of Bordeaux and Marseille.

Few remedies proved effective against the parasite, which fed off the vine roots and killed the plant. A few encouraging results were obtained by treating the vine with quicklime or carbon sulfide, or even by flooding the vineyard for months at a time to destroy the phylloxera eggs. The best solution in the long run, however, was to uproot the infected grapevines and replace them with stock native to North America, which was naturally resistant to the pest. The French wine-grape species were then grafted onto the American rootstock, where it could grow anew.

For Sancerre, this was a blessing in disguise. Since the winemakers were forced to replant their vineyard, they decided to experiment and graft mostly Sauvignon Blanc grapes onto the new stock instead of Pinot Noir. The switch was an instant success. Perfectly adapted to the climate and terroir, planted in light and porous soil that heats up rapidly, the Sauvignon grape propelled Sancerre wine to new heights, winning for this fruity, dry white an official appellation in the classification of French wines as early as 1936.

Fortunately for lovers of red wine, Pinot Noir did not disappear completely. Full of fruit and character, Sancerre red and rosé wines obtained their own appellations in 1959. Today, 2,200 hectares (5,500 acres) are planted with Sauvignon Blanc and 600 hectares (1,500 acres) with Pinot Noir, which translates to 79 percent of the vineyard given to producing white Sancerre, 14 percent to red, and 7 percent to rosé (fig. 6.1).

Spread over fourteen townships, the vineyard is partitioned among four hundred winemakers. Since the lots are far from uniform with respect to microclimate and soil, they deliver wines of unequal quality. In addition, the reputation and salability of the Sancerre label tempt many a winemaker to favor high yield and quantity over quality. If success has its pitfalls and can lead to somewhat standardized results, most Sancerre wines display an astonishing variety of flavors—not only from one year to the next but most noticeably across contrasting terroirs, which encompass

Figure 6.1 The vineyards of Sancerre, looking east from the hill of Chavignol, with the Sancerre hill in the background.

at least three soil types. Sancerre winemakers can keep the wines from each terroir separate to preserve their character, or blend them to fuse their complementary qualities and mitigate individual shortcomings.

The three distinct soil profiles derive from a wide range of rock strata, unearthed by erosion at different depths in the sedimentary pile — a sampling process made all the more complex by a set of faults that cut through the landscape, raising or lowering neighboring blocks and juxtaposing layers from different geological periods. Hence the grapevines of Sancerre tap limestone and clays from the Late Jurassic, but also younger Cretaceous and Tertiary strata.

A ROLLING LANDSCAPE

A traveler reaching Sancerre from the Loire valley to the east (via routes D 955 or D 920) faces a barrier of hills: the Sancerre butte in the middle, Saint-Satur to the north, and Thauvenay to the south. All three reach altitudes between 300 and 350 meters (1,000 and 1,160 feet), a relief of about 100 meters (330 feet) above the river valley.

The north–south-trending crest line reveals one of the major faults —

actually a bunch of parallel faults—that slice through the Earth's crust in the area: a weakness zone that dates to the early days of France's tectonic history, and was recently reawakened by extensional forces connected to the Alpine uplift farther east. This crustal extension has triggered the slipping and downward motion of the blocks east of the fault lines in a stepwise fashion, with the highest step making up the crest line and the lowest step the river valley (fig. 6.2). The Loire, like many a river, used this trough to trace its course—in this case to the north past Sancerre, before veering west toward Orléans.

On their eastern front, the Thauvenay, Sancerre, and Saint-Satur hills slope steeply down to the valley floor. Their western side, away from the river, slopes more gently down to a broad basin. Here, the vineyards roll out in waves, up to an escarpment barring the distant horizon: a steep cuesta leading up to the Berry plateau.

Set between its uplifted borders, the limestone bowl of the Sancerre vineyard has a low, hummocky floor, with small brooks running out to the Loire and cutting clefts between the hills of Sancerre and Saint-Satur (*cluse de la Colette*) and farther north (*cluse de la Belaine*). Besides opening up passageways to the Loire, the streams break up the hill front into several segments, adding north- and south-facing slopes to the major east- and west-facing ones. As for the hummocky basin floor itself, its chaotic nature, comparable to swells on the surface of an ocean, allows every possible orientation on a smaller scale between Sancerre, Chavignol, and Amigny.

The Sancerre vineyard also derives a clear climatic advantage from its topographical setting. To the west, the high Berry plateau ramps up before reaching its crest line and sloping down to the Sancerre basin. Rising along the ramp, western winds submit their damp oceanic air to a cooling trend that favors precipitation along the way; after dumping its water on the plateau, the air is drier and healthier when it blows over the vineyards. In the opposite direction, the harsh continental climate impinging from the east is somewhat blunted by the temperate influence of the Loire River. In the early fall, the morning fog that rolls in from the valley and up through the clefts spills over the vineyard, cooling the grapes and slowing their ripening—a prolonged maturation that promotes the development of stronger flavors in the grape juice.

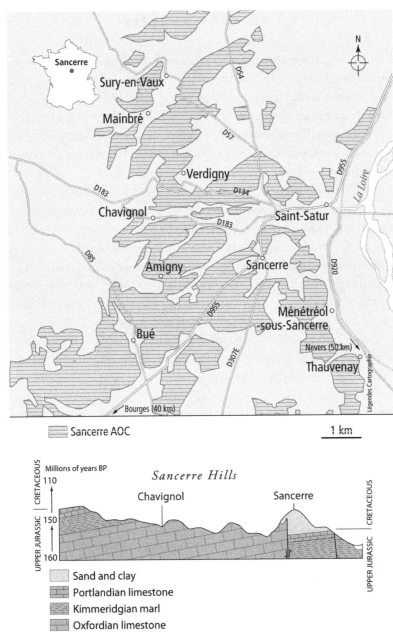

Sancerre

Sury-en-Vaux

Mainbré

Verdigny

Chavignol

Saint-Satur

Amigny

Sancerre

Ménétréol
-sous-Sancerre

Bué

Nevers (50 km)

Thauvenay

Bourges (40 km)

Sancerre AOC

1 km

Légendes Cartographie

Millions of years BP

Sancerre Hills

CRETACEOUS

110

Chavignol

Sancerre

CRETACEOUS

UPPER JURASSIC

150

160

UPPER JURASSIC

Sand and clay
Portlandian limestone
Kimmeridgian marl
Oxfordian limestone

Figure 6.2 *Top*, map of the Sancerre AOC vineyard. *Bottom*, geological cross section showing Jurassic strata interrupted by a fault zone at the level of the Sancerre hill (*right*), east of which Cretaceous and younger strata were down-dropped (on the Loire valley side). *BP* signifies before present. (Map by Legendes Cartographie; modified from a map by the Bureau interprofessionnel des vins du Centre, based on an Institut Géographique National document produced by Prisme Image et Communication.)

LIME, CLAY, AND FLINT

Sancerre wines owe some of their nuances to changes in slope and exposure across the vineyards. But their variety is enhanced by the succession
of distinct sedimentary strata. Hillslopes usually intersect several layers,
with the oldest at the bottom and the youngest at the top. Add to this a
regional trend: the floor of the Sancerre basin slopes upward toward the
west and the bounding cuesta, so that the ground level intersects increas-
ingly younger bedrock as it rises.

In the deeper parts of the basin, the vineyard is established on Jurassic limestone aged 160 million to 155 million years—an age or "stage"
called the Oxfordian by geologists, since the sediments were first mapped
around Oxford in England. In Sancerre's Berry region, the same-aged
sediment is referred to as "chalky Bourges limestone." This is the chronological stage we reached in the upper half of the Corton hill in Burgundy:
a time of tropical seas, when plankton and other marine animals showered the sea bottom with calcareous tests and shells.

Buried and turned to stone, then uplifted hundreds of meters above
sea level by tectonic forces, these calcareous sediments were ultimately
broken down by erosion. Water seeped into cracks, and ice wedges broke
up the layers into white platelets locally known as *caillottes* (fig. 6.3). These
limestone chips are found throughout the lower levels of the vineyard,
and the larger pieces are often used to pave the pathways. Where the
limestone layers are particularly thin, the broken-up chips are smaller and
called *griottes*.

Limestone *caillottes* and *griottes* heat up slowly in the sun, and after
dusk radiate their stored-up heat back to the soil and vines. This ground-
level microclimate causes the grapes to mature slightly faster than on other
types of soil, yielding light and fruity wines with a fresh bite. These wines
are highly representative of the Sancerre appellation as a whole, since over
half the vineyard is planted on *caillottes* and *griottes*. Raised on limestone,
they are to be drunk young, in their second or third year.

The Jurassic sediment changes nature and character as you go up one
level, either by climbing a hill or by following the gradual trend in elevation of the terroir westward toward the bounding cuesta. From the Oxfordian, one reaches the Kimmeridgian age of 155 million to 150 million

Figure 6.3 Sancerre's Jurassic limestone is broken up into thin slabs, locally called *caillottes*, by freeze-thaw action.

years ago, during which the sediment records sharp fluctuations in sea level. Algae, plankton, and seashell debris are cemented together in an entirely calcareous matrix (a limestone) or else have a little clay mixed in (a marl), when the water was deeper and calmer.

The marls of Sancerre contain ammonite fossils—those great mollusks that lived in coiled shells and cruised the open waters. We encountered older species of the group in Pouilly-Fuissé and in Burgundy. Among the new "models" found at Sancerre, specialists recognize species introduced from the northern Baltic ocean as well as the southern Tethys Sea—proof that the shallow sea covering France at the time was connected to both.

The Kimmeridgian marls and their fossils outcrop at the top of the hillslopes, for instance at Chavignol on La Côte des Monts Damnés (literally, "The Hillslope of the Damned Mountains") and in the westernmost part of the vineyard, up the regional slope, where they underlie the Berry plateau and feed the neighboring vineyard of Menetou-Salon (where the wines are comparable in character to those of Sancerre).

Kimmeridgian marls break down to yield clay-rich soils that become very sticky when wet: they are nicknamed *terres amoureuses* ("soils in love") because they cling relentlessly to shoes. Light brown to bluish in color when damp, the marls turn white when baked in the sun, so they are locally called *terres blanches* ("white soils"). Carrying less pebbles and platelets capable of storing solar heat, Kimmeridgian marl is considered a "cold" type of soil, with the grapes maturing significantly slower than on limestone *caillottes*. This slow maturation profile actually gives more time for the grapes to concentrate aromatic phenols in their juice, which brings complexity to the wine of that terroir.

Sancerre wines from the Kimmeridgian marls are thus known to be rich and full bodied, with a floral and earthy bouquet. They mature more slowly than the fresh and fruity limestone Sancerres, and need about three years to reach their prime.

Besides these two terroirs—limestone *caillottes* and white marls—there is a third type of terroir at Sancerre, on the eastern front. This terroir straddles the fault zone and contains younger sedimentary layers lowered to ground level by the down-dropping of the rock pile. The hill of Sancerre actually belongs to a down-dropped block that brought its Cretaceous cover at the same level as the Jurassic terroir, west of the fault. So while vines on the western side of Sancerre grow on Jurassic 150-million-year-old limestone and marl, vines on the eastern side feed on Cenomanian limestone (mid-Cretaceous) that is "only" 100 million years old. There is an even younger rock cover (a 50-million-year-old Eocene conglomerate) that caps the top of the hill and contributes rubble to the vines below.

The eastern, sun-facing slope of the Sancerre hill is therefore very different from the western slope. It is naturally steep, its soil rich in flint, which provides quite a different flavor to its wines. The stony soil heats up the vines and makes for firm wines with a bite, a bouquet of broom and acacia flowers, and above all the strongly reinforced gun-flint aroma of Sauvignon Blanc.

Limestone, marl, and flint: this diversity of terroirs allows the winemaker to bottle distinct Sancerre types, from fruity to smoky, rich, and full bodied. Another approach consists in blending grapes from different lots, to merge their qualities. You will find both winemaking philosophies at Sancerre, and a visit to a couple of estates, among the four hundred or so winegrowers sharing the appellation, brings out some good examples.

POUILLY FUMÉ AND POUILLY-SUR-LOIRE

Right across the Loire from the Sancerre district, on the river's right bank, the wines around the village of Pouilly-sur-Loire, namely the appellation Pouilly Fumé, are much lesser known than their big-name neighbor, but hard to tell apart and basically just as good.

There is a confusion to be avoided here. A completely different wine in Burgundy possesses the same name root of *Pouilly* and is called Pouilly-Fuissé (see chapter 4). It is based on Chardonnay. Here in the Loire valley, Pouilly Fumé and Pouilly-sur-Loire are completely different wines, made of completely different grapes.

The Pouilly Fumé terroir is symmetric to that of Sancerre across the Loire valley, with river-facing Cretaceous and Eocene slopes, inward of the normal faults, and Jurassic strata jutting outward into the flanking basin. As in Sancerre, the Oxfordian limestone makes the transition to Kimmeridgian marl, and the wine's character varies accordingly. The marl is particularly evident up the hill of the most prominent knoll, named Saint-Andelin, and the Oxfordian limestone is mainly found on lower slopes to the south.

Competing with Sancerre, the district of Pouilly-sur-Loire raises Sauvignon Blanc: an appellation known as Pouilly Fumé because of the natural, smoky character of the grape. But the district also produces a very local wine, called Pouilly-sur-Loire. Hiding behind its village name is a very old and forgotten grape variety, Chasselas, which yields a light white wine with relatively low alcohol content. Only mildly acidic, it is a good match for pan-fried minnows and other small river fish.

THE VINES OF CHAVIGNOL

The tiny village of Chavignol, with a population of two hundred souls, is mostly known for its excellent goat cheese—*crottin de Chavignol*—one of the first cheeses in France to receive an Appellation of Origin, a recognition of terroir similar to that for wine (there are now forty-five AOC [Appellation of Controlled Origin] cheeses in France). But despite the huge output of Chavignol goat cheese—close to 20 million *crottins* a year—you will not see many goats in the village. They have been pushed back to the distant plateau beyond the cuesta, for the town is at the very heart of the Sancerre winegrowing district, and goats don't stand much of a chance against Sauvignon Blanc. Try the Chavignol goat cheese if you have a chance, however. It is excellent, and a perfect match for a glass of white Sancerre.

Today, Chavignol is one of the leading producers of Sancerre wine, and both large and small estates operate out of the village. One of the largest estates is the Henri Bourgeois domaine, which operates its winery in the upper part of town. From the terrace above the wine vats, looking east, you get a view over the rooftops, the vineyards, and the Sancerre hill in the background. To the north and west, the nearby cuesta in back of the village takes the shape of two steep hills that deliver excellent Sancerre Crus: La Côte des Monts Damnés and Le Cul de Beaujeu.

In the fall, the narrow streets of Chavignol are full of the smell of freshly pressed grapes, and it was on one of those October mornings that I met up with Raymond Bourgeois, one of the three family members running the estate. The timing was ideal, since the Bourgeois staff had dug trenches in all their "climates," to view and study the soil in cross section. They were also collecting spadefuls of soil to frame in glass cases back at the visitor center, to illustrate the makeup of the Sancerre terroir.

Raymond sent me out in the field with staff member Yannick, who guided me to all the places where trenches had been dug—they were to be filled in right after our visit. We started off our trek at a young siliceous terroir on the southern slope of Saint-Satur hill, near the cleft connecting the Sancerre basin and the Loire valley. This "climate," named Les Ruchons, showed a lot of clear-colored flint set in a pinkish soil that turned brownish orange at depth. The names of some of the climates here are colorful evocations of the soil's stony makeup, be they Gratte-Sabots (the "Clogs-Scraper") or Chailloux-Tue-Chien (the "Dog-Killing-Stones").

Flint in the soil translates into spicy, smoky aromas: wines best described as strong and sharp. They are rather austere and mineral in their youth, but gather complexity with time. We tasted the estate's "flint" Sancerre when we returned to the winery later that morning, and Raymond Bourgeois described its mineral character as "a decoction of pebbles," pointing out as well its nuances of candied fruit and spices.

Leaving the cleft of Saint-Satur and its rather young flint, Yannick and I next headed into the vineyard basin, crossing the invisible fault line into the Jurassic period. The second trench was excavated in marine limestone—the fragmented *caillottes* and *griottes* terroir—on a rather steep, south-facing slope, in a climate known as Les Bouffants. Here lie chunks of limestone of different sizes, including larger, whitish *caillottes* and smaller, worn-down, yellowish *griottes*. At the Bourgeois domaine, grapes from these calcareous lots are generally blended with those from the marly, clay-rich *terres blanches*, to obtain a balanced wine. Tasting the

blend at the winery, you can pick out the fresh, fruity touch of the limestone component, and the powerful contribution of the marl-raised wine,
adding aromas of citrus fruit and pineapple.

Besides Sauvignon Blanc, the *caillottes* and *griottes* limestone terroir
also welcomes Pinot Noir—red wine accounts for 14 percent of Sancerre's production—and, as is the case with white wine, the output from
these calcareous lots is blended with Pinot Noir from marly lots in order
to obtain a balanced red that gives off aromas of red berries and morello
"griotte" cherry in particular. Incidentally, the fact that the tart cherry
type is named "griotte," like the limestone pebbles associated with the
wine, is a sheer coincidence.*

THE HILL OF THE DAMNED

Above the Oxfordian limestone *caillottes* and *griottes*, we enter the Kimmeridgian age of 155 million years ago, when marine sediments on the
Sancerre site turned to clay-rich marls—a terroir known today as the
terres blanches. One of the best examples is the hill overlooking Chavignol, which has a south-facing, steep slope and a reputation of excellence:
La Côte des Monts Damnés (fig. 6.4).

Climbing the steep hill, which affords a superb view of the village and
its small cemetery surrounded by vines, we reach the Bourgeois lots at
midslope and the trench dug into marls. Light gray at the surface, the soil
turns to a deeper gray at depth, with bluish streaks.

Such great soil and exposure yield superb Sancerre wines at Les Monts
Damnés. When young, they are fine and citric, with notes of grapefruit.
But with the magical touch of their marly terroir, three or four years of
aging will bring out rich and complex aromas, evoking candied fruit.

Walking up the slope, you get a sense of the variations in sea level as the
Earth headed toward the end of the Jurassic period. At the foot of the hill
we find shell limestone, with some beach cobbles mixed in—the sign of
coastal shallow waters. Perhaps the site of Sancerre had emerged from the
Jurassic sea at that time, and was pounded by the surf.

The next layers above the coastal part are the famous clay-rich marls,

---

* The word *griotte* meaning "pebble" comes from *crais* (*craie*), meaning "chalky limestone,"
whereas the word *griotte* that describes the cherry variety derives from the Catalan word *agre*,
which means "sour."

Figure 6.4 The renowned hillslope of Monts Damnés, made up of Jurassic marl rich in oyster fossils, overlooks the village of Chavignol.

indicating that the site once had been submerged in deeper, calmer waters. Either the global sea level had risen at the time, or else the Sancerre seafloor had subsided, beneath the weight of its growing pile of sediment.

The conditions appear to be cyclical: benches of limestone alternate with layers of marl as we climb the slope and the geological timescale, as if the depth of the sea had varied periodically. Maybe the seafloor had subsided in a series of steps, each time the weight of the growing sediment reached a threshold.

In these accumulating sea-bottom oozes now turned to marl, we find ammonite fossils and even occasional vertebrae from marine reptiles that once cruised the Sancerre seas. The nearby marls of the Berry plateau have even delivered the nearly complete skeleton of a plesiosaur—a Loch Ness–like, long-necked monster. This particular specimen is 4 meters (12 feet) long and 150 million years old, and is on display at the National History Museum in Bourges.

On Les Monts Damnés we also run across clam fossils, or at least their Jurassic ancestors named *Pholadomia*, which dug a burrow in the sea bottom. You can find specimens as large as a fist on the hillslope, and in keeping with the hellish reputation of the place, where one might well die working, these fossils are nicknamed "winegrowers' hearts" (*cœurs de vignerons*).

This Kimmeridgian marl, full of fossils and legends, ends at the top of the hill with a layer chock-full of oyster fossils. The shells are small and shaped like hooks or commas, deserving for that reason their species name of *Ostrea virgula* (the Latin word *virgula* means "small twig" and, by extension, a comma). The oysters lived in large colonies on the sea-floor; their shells are cemented together along adjacent valves and present

---

CHABLIS

Chablis is not a wine appellation of the Loire valley but instead belongs to the watershed of the Seine River. East of the city of Auxerre, the vineyard is established on the banks of the Serein, confluent of the Yonne (itself a confluent of the Seine). It is mentioned in this chapter, because its terroir is similar to that of the Sancerrois. At Chablis, the vines are also planted on marly calcareous soil of Jurassic Kimmeridgian stage, with the same oyster fossils (*Exogyra virgula*) as at Sancerre.

The big difference is that at Chablis the grape variety is Chardonnay rather than Sauvignon, and the vineyard is part of Burgundy (actually the northernmost appellation of Burgundy), flanked by the Auxerrois appellation to the southwest and the Tonnerrois appellation to the northeast.

These differences make Chablis a unique white wine that marries the Burgundy style of Chardonnay with the oyster marl terroir found in Sancerre's Kimmeridgian lots. Add to this a rather rough climate, with severe frosts in winter and spring. Chablis is famous for its bouquet of brioche bread and hazelnut, its body, and its minerality. Some wine tasters also think they taste the iodine aroma of the long-deceased oysters and their terroir (or is it present-day oysters, often served with Chablis, that fool their senses?).

The Chablis vineyard is shaped like a butterfly, with each wing covering a bank of the Serein River. You can rent a canoe or kayak and glide down the Serein's lazy current to view the vineyards while the river snakes its way down the valley, approaching at times the western slopes, and at other times the eastern ones.

At the top of the slopes, the rim of the cuesta is a layer of hard limestone belonging to the final Portlandian stage of the Jurassic period. Chardonnay vine is also grown up on that plateau, and yields the separate appellation of Petit-Chablis: lighter, fruitier wine, but with the same distinct minerality as Chablis itself.

Chablis achieves excellence on the best-exposed Kimmeridgian slopes, claiming Premier Cru status on both the left and the right banks. Exposed southwest to southeast, the right bank possesses a particularly steep slope right above the village of Chablis, where seven climates are classified as Burgundy Grands Crus. Their aromas span a wide palette that ranges from honey and cinnamon to citrus fruit, flowers, and smoky, earthy notes. Wines with the fullest body come from the lower slopes, like in the Grenouilles climate, and the finest wines from the higher slopes, like the top lots of Les Clos and Vaudésir.

a dense, packed structure, winning them the name honeycomb oysters. You need only break up a handful of marl in the palm of your hand to collect dozens of them, each the size of a fingernail. Some 150 million years of evolution separate these Jurassic oysters from the ones we eat today with a glass of Sancerre.

At Les Monts Damnés, the oyster bench marks the end of the great marl terroir and the end of the reign of Sauvignon Blanc. The top layers of the Kimmeridgian do not have enough clay to support white grapes, and so are planted instead with Pinot Noir, which does very well in low-clay conditions, thereby earning a spot on the hill to produce Sancerre red.

A few more meters and we reach the final stage of the Jurassic, the Tithonian stage (or Portlandian stage in Europe) of 151 million to 145 million years ago. It was an age of hard limestone that stands out here as a resistant ledge rimming the cuesta. In turn, the cuesta underpins the rolling surface of the Berry plateau, its wheat fields, Menetou-Salon vineyards, and pastures roamed by Chavignol goats.

THE CLOSE OF THE JURASSIC

The Portlandian limestones herald the end of the Jurassic period, a long and rather stable age marked by continental drift and an overall rise in sea level. It saw the proliferation of dinosaurs on land and of great marine reptiles in the sea.

The end of the period coincides with a transient drop in sea level; consequently, the waters retreated from the Sancerre area. Only the deepest basins in France kept enough water to continue recording in their sediments and fossils the closing days of the Jurassic.

At Sancerre, the submarine landscape and its cover of sediment emerged after 50 million years underwater. This new land became covered with tropical forests and was probably roamed by herds of End-Jurassic dinosaurs. However, there is a lack of fossils from this period at Sancerre, since no sedimentation took place here at that time.

What we do know, based on fossils preserved elsewhere in the world, is that the ecosystem was shaken down 145 million years ago. The new age to follow—the Cretaceous period—did not feature the same animal populations as before, both on land and in the sea. Many genera of dinosaurs entirely disappeared, especially the larger-sized species such as America's famous *Diplodocus*—30 meters (100 feet) long from head to tail, and with a brain no larger than a 60-gram (2-ounce) *Crottin de Chavignol* cheese.

In the marine realm, many families of ammonites and bivalves disappeared as well, as did several families of tortoises and other marine reptiles, including an entire family of ichthyosaurs—the dolphin-shaped, ammonite-eating predators.

The sea-level drop probably played a role in this End-Jurassic mass extinction, since it certainly shrunk the area of the marine shallow-water habitat. It does not explain, however, the equally noticeable dip in the terrestrial dinosaur diversity. Did the climate deteriorate on a planetary scale? Did the Earth suffer the impact of a large asteroid, as it later would at the end of the Cretaceous period? There are indeed several impact craters around the globe that coincide with the end of the Jurassic period, namely Morokweng (a 70-kilometer [44-mile]-wide crater) in Africa and Mjølnir (40 kilometers [25 miles]) in the Barents Sea, north of Sweden. The impacts that created those craters were not that huge, so the reason for the extinction could be completely different—but the coincidence is worth noting.

The next events, at the start of the Cretaceous period, are registered again at Sancerre, thanks to a switch back to a sea-level rise, and a new batch of sediment accumulating and entombing fossils in the French basins. This chapter of Cretaceous evolution is best engraved in the sediments of the lower Loire valley and in Provence—the next regions we shall visit.

As for the Sancerre region, its land remained above water for a while. Sedimentation returned in earnest to the area 130 million years ago, after about a 15-million-year gap in Early Cretaceous layering action.

The first Cretaceous strata deposited here are flush with the vineyards on the eastern slope of the hill of Sancerre, where they were brought down by fault motion. As already mentioned, this pass of Cretaceous sediment is capped by a much younger lid of chert and flint conglomerate rock, only 50 million years old, that sheds its siliceous chunks and pebbles down the slope.

The flint layer brings to this batch of Sancerre wines a mineral, smoky bouquet (fig. 6.5). This character, true to Sauvignon, is enhanced around the hills capped by the siliceous rock. Thus, we shall visit the vine-growing hill of Sancerre itself, where many estates own lots. Among these wine-

Figure 6.5 At the foot of the Sancerre hill, the terroir is enriched with flint shed from the upper slopes, which contributes to the wine's smoky bouquet.

growers is one of the smallest domaines in town, which nonetheless turns out two contrasting cuvées of Sancerre: the domaine of Vincent Grall.

SANCERRE AND FLINT

Born in Brittany and later moving to Sancerre, Vincent Grall learned the wine trade from his uncles before starting his own vineyard. He owns about 3 hectares (7½ acres) of Sauvignon Blanc distributed on five lots around the hill of Sancerre: two on Le Plateau, which makes up the south-western foot of the hill where it grades into the Jurassic basin; and three on the southeastern, much younger Cretaceous slope facing the Loire valley, in a climate named Manoir de l'Étang ("Pond Manor").

I met Vincent in the little winery where he vinifies and bottles his Sancerre, on the street running alongside the ramparts of the fortified city and named *rempart des abreuvoirs* ("rampart of the water places"), although there is more wine than water to be had here today. My visit was in mid-October and the harvest was in for the year, the grapes picked by hand and brought unbruised as much as possible to the winepress. Vincent was now overseeing the fermentation process, checking his vats and sampling their juice with his pipette, to measure temperature and density (the latter a proxy for alcohol content). The morning of my visit, the vat temperature was 26°C (78.8°F), and the juice density had just fallen below 1, reading .995. Alcohol was definitely taking over.

The winemaker was also sampling the yeast deposit accumulating at the bottom of the vat. Every fortnight, according to the phases of the moon (since Vincent believes in the exotic craft of biodynamics), the fermenting wine is stirred—an operation called *remuage* or *battage* in French—so that the yeast deposit is resuspended in the wine and transfers more aromatic molecules to it. In January, the wine will be moved to another vat to rest before being bottled in March.

Vincent blends the wine from both his climates—Le Plateau and Manoir de l'Étang—in order to reach a carefully sought balance. He makes his Tradition cuvée mostly with grapes from his Plateau limestone lots. His more expensive Le Manoir cuvée, on the other hand, is made mostly from the eastern clay-rich lots of Manoir de l'Étang, with a little limestone Plateau wine thrown in for balance.

It is late afternoon when we leave the winery to check out the terroir in the light of the setting sun. Le Plateau is a very level strip of land, with the slope of the Sancerre hill a quarter mile away to the north and the twin hills of Thauvenay and Orme-aux-Loups to the east. According to the geological map, Vincent's lots are right on the fault zone that separates Jurassic limestone on one side from Cretaceous sand and flint on the other. The soil is brown—a caramel brown with a cinnamon hue—stuffed with nuggets of flint. We also pick out blocks of conglomerate stone, from the Eocene concrete-like cap rock, that was shed off the hilltops.

This flint scree over Jurassic limestone gives the Plateau wines a natural balance between fruit and smoke, which Vincent Grall captures in his Tradition Sancerre. Wine tasters describe it as having an explosive and tonic bouquet, aromas of grapefruit, and a fresh, mineral palate.

On the opposite, eastern side of the hill, Vincent's three other lots produce a noticeably different wine. Here we stand on coastal-water Cretaceous sands and clays deposited in Albian times, 110 million to 100 million years ago. Vincent's lots are right above a hollow containing a manor next to a pond, hence its name Manoir de l'Étang ("Pond Manor"). The sand and clay make up a brownish soil, with again some flint shed off the Sancerre hilltop. This eastern, Loire-facing terroir benefits from the

morning sunlight and from the microclimatic influence of the river, treating the winemaker to a richer, more complex palette than the wines coming from Le Plateau on the western slope. It has a more mineral, vanilla-
rich fragrance, with a touch of roasted almond and a floral bouquet of
peach and orange blossoms.

After climbing the last stages of the Jurassic, we end our tour of the
Sancerre vineyard in the early strata of the Cretaceous period. To get to
the heart of this new geological chapter and its most representative wines,
we now drive down—or sail down—the Loire valley toward Tours, Saumur, and Chinon.

CHAPTER SEVEN

ℭhe ℭentral Loire Valley

Bourgueil, Chinon, and Saumur

In its upper reaches, the Loire valley celebrates Sauvignon Blanc, the grape of Sancerre, Pouilly-Fumé, and other dry, elegant, fruity whites. As we proceed down the river, where the Loire meets two of its major confluents—the Indres and the Vienne—the broadened valley is home instead to Cabernet Franc and red appellations: Bourgueil, Saint-Nicolas-de-Bourgueil, Chinon, and Saumur-Champigny (fig. 7.1).

All four appellations share an exceptionally mild climate, bringing together outstanding terroir that mixes Cretaceous limestone with sand and gravel rolled in by the rivers. On this terroir and in this climate, Cabernet Franc is the king grape, delivering excellent red and rosé wines. There are also a few lots of Chenin Blanc providing some white, but Chenin truly comes into its own farther downstream at Savennières and Côteaux-du-Layon (see chapter 1).

The red Cabernet Franc wines of the Loire valley are famous for their lightness and fruity aromas of cherry, black currant, and raspberry. But they come in two different styles. On the one hand, there are those light, fruity, and dry "thirst-quenching" wines that are a perfect match for summer barbecues. On the other hand, you can find richer varieties that age remarkably well and yield powerful, well-built red wines. This dual personality of Cabernet Franc in the Loire valley comes from two contrasting terroirs running parallel to the river: the gravel terraces, mixed with clay, that yield the lighter, fruitier *vins de gravier* ("gravel wines"); and the strata higher up of chalky limestone that yield the more complex *vins de tuffeau* (fig. 7.2).

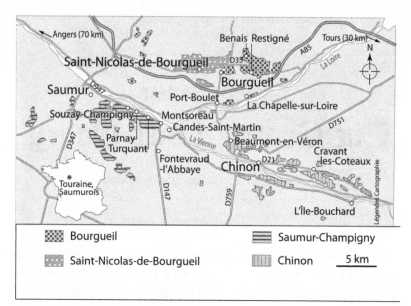

Figure 7.1 The vineyards of Bourgueil AOC, Saint-Nicolas-de-Bourgueil AOC, Chinon AOC, and Saumur-Champigny AOC. (Map by Legendes Cartographie.)

BOURGUEIL AOC

Region:	Loire valley (Touraine)
Wine type:	red, rosé
Grape variety:	Cabernet Franc
Area of vineyard:	1,400 hectares (3,500 acres)
Production:	70,000 hectoliters/year (9,310,000 bottles/year)
Crus:	none, but reputed "climates" (e.g. Grands Champs)
Nature of soil:	calcareous, or gravel
Nature of bedrock:	chalky limestone (*tuffeau*)
Age of bedrock:	Cretaceous, Turonian stage (90 million years ago)
Aging potential:	3 to 10 years (*tuffeau* cuvées)
Serving temperature:	14–15°C (57–59°F) (gravel-raised red)
	16–17°C (61–63°F) (*tuffeau*-raised red)
To be served with:	veal, rabbit, poultry, tajine

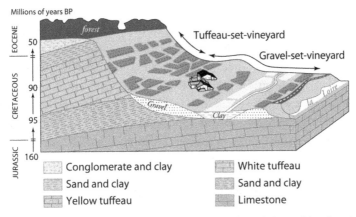

Figure 7.2 Cross section of the Bourgueil AOC vineyard, showing gravel alluvium at the bottom of the valley and strata of Cretaceous *tuffeau* limestone at midslope, which yield two different-tasting wines. *BP* signifies before present. (Map by Legendes Cartographie; modified from the map of the Syndicat des vignerons de l'AOC Bourgueil.)

The *tuffeau* in question is a white, chalky limestone sprinkled with glistening flakes of mica, which is the founding rock of the region in more ways than one. Light, porous, and permeable, it readily drains the water off the slopes and serves the vineyard well. *Tuffeau* also qualifies as a superb building stone that has been a favorite of masons and sculptors for centuries. Churches and castles have been built with it, including the great châteaux of the Loire: Amboise, Azay-le-Rideau, Blois, Chambord, Chenonceaux and so many others.

As for the numerous quarries dug into the hillslopes, many were converted to underground "troglodyte" habitats—the main style of living along the Loire valley as recently as the nineteenth century. Some of the galleries are still used today as offbeat dwellings, and others have been converted into mushroom farms and, of course, wine cellars.

Vine growing in the central Loire valley dates to at least Greco-Roman times. In his *Naturalis Historia*, naturalist and philosopher Pliny the Elder (AD 25–79) wrote about a successful grape variety in the Loire valley, *Vitis biturica*, which could well be the ancestor of Cabernet Franc. Pliny describes it as "a quality plant that stands the cold weather, storms and rain, and yields a wine that can be kept for a long time and improves over the years."

But despite this early start, the Touraine region truly took off as a major wine-producing region in the Middle Ages, when the Abbey of Bourgueil was founded in AD 990 and took over the vineyards' destiny.

The Benedictine abbey, 50 kilometers (30 miles) downstream of Tours, is strategically positioned on the old Roman road that links Tours to Angers. It is also located downstream from the confluence of the Loire and the Indre Rivers, where the valley enlarges and where there is lots of land for agriculture. The abbey itself was known to host a rich herbal garden, with exotic varieties such as star anise, coriander, and licorice, and it watched over the vines and the winemaking. As early as 1089, the abbey's prior, Baudri de Bourgueil, proudly promised his visitors "a good wine in stock."

BOURGUEIL AND THE POET

The Abbey of Bourgueil long ruled over the region, both agriculturally and politically. Henry II Plantagenet (1133–1189), ruler of France and England and founder of a long-lasting dynasty, held the Estates General of his provinces here in 1156, and Pope Innocent III (1160–1216) guaranteed the abbey support and protection in 1208. Many French kings and queens spent time at Bourgueil, drank the local wine, and surely made the right decisions for their kingdom.

Today, we can still admire the refectory's impressive vaulted ceiling; an elegant, suspended stairway ascending the walls with no supporting pillars; and around the buildings, "suspended" terraced gardens that were added during the Renaissance. Poet Pierre de Ronsard (1524–1585), close to the Royal House and to King Henri II of France, was particularly fond of Bourgueil. An epicurean, Ronsard was famous for praising love, wine, and the fleeting moment. In book 2 of his *Odes*, he proclaims:

> I live not for tomorrow
> Page, pour into my cup again,
> Fill this great glass to the brim . . .
> For the brain is never quite sane
> Unless it is quenched with love and wine.

Ronsard had good reason to drink. In 1555, he fell helplessly in love with fifteen-year-old Marie Dupin, who lived outside Bourgueil in the nearby village of Port-Guyet. His love letters—*Lettres à Marie*—are among his most passionate poems. Marie's house (her father was an innkeeper) still stands, on the edge of the vineyard.

Bourgueil and Port-Guyet are built on top of an ancient alluvial terrace, about 15 meters (50 feet) above the Loire's present riverbed. Although there are some oases of vines planted on gravel bars at lower elevations, most of the Bourgueil vineyard is set on this upper pebbly terrace and on the chalky bedrock higher yet on the hillslope, straight up to the tree line.

The Bourgueil vineyard enjoys a gentle climate, typical of the Loire valley. Spring comes early, summers are hot and sunny, and ocean air from the Atlantic blows up the valley, taking the edge off the summer heat. The forest on the hillcrest also protects the vines from the north wind.

We get a sense of this microclimate as soon as we reach the Bourgueillois (Bourgueil country) via route D 952, following the levee along the Loire valley. From its perch, the road overlooks quaint gardens planted with magnolia trees, palm trees, and other tropical species.

The road passes through Chapelle-sur-Loire, built on the riverside and surrounded by sunflower fields, greenhouses, pear-tree orchards, ponds stocked with carp and pike, and vine rows of Cabernet that are planted on gravel mounds locally known as *puys* or *montilles*. This alluvial terroir—a hodgepodge of sand and gravel rolled in by the Loire and its confluents, with some finer clay mixed in—yields the typical *vin de gravier* ("gravel wine"): a light and fruity red, almost purple in color, giving off aromas of red and black berries.

SAINT-NICOLAS-DE-BOURGUEIL

The appellation Saint-Nicolas-de-Bourgueil is the continuation of the Bourgueil vineyard westward, centered on its namesake village. Only 3 kilometers (2 miles) separate the villages of Bourgueil and Saint-Nicolas, and both share the same hillside overlooking the Loire valley. Nonetheless, the residents of Saint-Nicolas were able to claim a separate appellation for their wine, which involves 1,000 hectares (2,500 acres) of Cabernet Franc (against 1,400 hectares [3,500 acres] for their neighbor).

Although climate, elevation, and exposure are roughly the same for both villages, their soil distribution is slightly different. There is less *tuffeau* terroir (30 percent) and more gravel (70 percent) at Saint-Nicolas. There is also an effort there to standardize its wine. One-third of the village's output is produced by the winemakers' cooperative, with the emphasis placed on the wine's light and fruity character. Even legislation favors this lightness, since higher yields are authorized for Saint-Nicolas-de-Bourgueil (60 hectoliters per hectare [3.2 tons/acre], versus 55 [2.9 tons/acre] for Bourgueil). As a result, Saint-Nicolas wines also have less of an aging potential: under five years, whereas a good *tuffeau* Bourgueil can easily be kept for ten.

At Chapelle-sur-Loire, we take the side road uphill toward the villages of Restigné and Benais, and on we drive through the higher terraces of older alluvium. These also yield light and fruity gravel wines, so typical of Bourgueil and of Loire reds as a whole.

The flip side of Bourgueil—a completely different terroir—stands out when we reach the upslope villages and start noticing outcrops of white, chalky limestone alongside the road and among the vines. This is the local bedrock.

TUFFEAU WINE

In the Touraine region between Tours and Angers, the Loire dug its course into a limestone plateau of marine sediments about 100 million years old (mid-Cretaceous), laid down at the bottom of a tropical sea that covered the northern half of France under several dozen meters of water.

Grapevines love this limestone *tuffeau*, a rock variety not to be confused with tufa (an alkaline, lacustrine limestone) or tuff (a volcanic ash). *Tuffeau* is a marine limestone—basically a calcium and magnesium carbonate—but it also includes mineral particles worn off the nearby crystalline massifs of Brittany and central France and dumped by the rivers into the shallow sea: flakes of mica, tiny quartz grains, and iridescent opals. They give the limestone a stucco-like sparkle.

Cabernet Franc enjoys this type of substrate. It finds better, deeper rooting in *tuffeau* than it does in gravel. The resulting wines have more punch and can age up to ten years, contrary to their gravel-raised cousins. In particular, the mica flakes in *tuffeau* break down into a green clay known as glauconite, which provides sodium, potassium, magnesium, and iron to the vines and helps fight off chlorosis, an iron deficiency that inhibits the production of chlorophyll in plants.

The village of Benais, like most villages on the upper slopes of Bourgueil, is built on *tuffeau* bedrock. Do not expect a sharp boundary, however, since the slopes are gentle and the mother rock hidden under the oldest, highest gravel bars of the Loire valley. There are no white cliffs jumping out at us, but under the thin soil cover, colored red by the iron oxides, the white limestone lurks. The wine cellars behind the houses are dug directly into it.

The Domaine de la Noiraie, on the outskirts of town, straddles the invisible line between gravel and *tuffeau*. Below the road, on the Loire valley side, its vines are rooted in gravel. Uphill from the road, its lots are established on *tuffeau*. Brothers Jean-Paul and Michel Delanoue, Michel's wife Pascale, and their son Vincent run the Noiraie estate, which provides a good example of the dual nature of the Bourgueil terroir.

Their gravel-raised Bourgueil red is light and fruity, as would be expected, and should be drunk one or two years after bottling. It has typical aromas of red berries, leaning toward raspberry.

If you are looking for *tuffeau*, however, you need not go very far. In the Noiraie domaine courtyard, a flight of steps leads down to the cellar, where oak barrels and stacks of bottles lie under a thin limestone ceiling, over which is only 2 meters (6 feet) of soil. This cross section shows us that the vines need pierce only a couple of meters of soil to reach the bedrock, and even less higher up on the plateau, where the Delanoue family owns several lots in one of the top Bourgueil climates: Grand Mont ("Big Hill").

Grand Mont lies a quarter mile outside town, marked by a signpost and a roadside parking lot alongside the entrance to an underground quarry—now the storage haven for the wine. Stroll down the ramp to the quarry, and you will enter a vast gallery where limestone blocks were extracted, the white ceiling pierced by skylights. Besides storing wine here, the Delanoue family uses the quarry to host wine-tasting parties and other special events, including concerts by their musician buddies, the Bourgueil Country Brass Band.

Tuffeau wine is full bodied and well structured, while at the same time preserving the fruity aroma and silky texture of Cabernet Franc. In contrast to its gravel-raised cousin that should be drunk young, *tuffeau*-raised Bourgueil should be given at least two or three years, if not more, to age in order to fully develop and express its aromas. More care is therefore given to the winemaking process: the grapes macerate for up to three weeks—one week longer than gravel wine—so as to extract more tannin from the grape skins.

Bourgueil winemakers often vinify their gravel and *tuffeau* Cabernet separately, but they can also blend them together, to fuse their qualities. For their Saint-Vincent brand, for instance, the Delanoue family spikes an essentially gravel-raised wine with a splash of its *tuffeau* cousin, thereby bringing structure and depth to the blend.

CHINON

While the vineyards of Bourgueil and Saint-Nicolas-de-Bourgueil occupy the right bank of the Loire valley, their cousin and competitor Chinon flaunts its easternmost rows of Cabernet Franc directly opposite, on the west bank. Most of its vineyard, however, spreads out on a vast alluvial terrace 10 kilometers (6 miles) downstream, where the Vienne River joins the Loire.

CHINON AOC

Region:	Loire valley (Touraine)
Wine type:	red, rosé
Grape variety:	Cabernet Franc
Area of vineyard:	2,400 hectares (6,000 acres)
Production:	113,000 hectoliters/year (1,500,000 bottles/year)
Crus:	none, but reputed "climates" (e.g. Le Clos de l'Écho)
Nature of soil:	calcareous or gravel
Nature of bedrock:	chalky limestone (*tuffeau*)
Age of bedrock:	Cretaceous, Turonian stage (90 million years ago)
Aging potential:	5 to 10 years (*tuffeau* cuvées)
Serving temperature:	14–15°C (57–59°F) (gravel-raised red)
	16–17°C (61–63°F) (*tuffeau*-raised red)
To be served with:	veal, rabbit, poultry, lamb, barbecue

To enter Chinon's heartland, we need to sail or drive around the spit formed by the merging rivers—a checkerboard of vines and cottages called Le Pays de Véron—and head up the Vienne valley to the stunning medieval village of Chinon. Erosion by the river has sculpted the edge of the limestone plateau into towering cliffs, crowned first by a Roman fortress, later by a medieval castle with ramparts, a moat and a drawbridge, and sturdy towers and dungeons.

The château of Chinon was occupied and further fortified by Henry II Plantagenet, Richard the Lionheart, and Philip Augustus, but it is best known for the legendary encounter in February 1429 between Charles VII (1403–1461), heir to the French throne, and seventeen-year-old Joan of Arc (1412–1431), who encouraged him to confront the English occupants north of the Loire valley, offered to raise an army, and predicted decisive victories over the enemy—a prophecy that would come true in less than six months' time.

Six centuries later, the Chinon castle is still standing. Two surviving towers, part of the royal lodgings where Charles VII met Joan of Arc, flank the drawbridge, and the dungeon where Joan stayed during her visit also remains.

To build the monumental castle, the architects and masons of the Middle Ages extracted the stone directly underfoot in the *tuffeau* plateau, digging shafts and galleries and hoisting the blocks right onto the building site. Riddled with galleries and sagging under the weight of castle walls and village houses, the top of the plateau caves in now and then, including a spectacular collapse that occurred east of the castle in 1921.

In the lower part of town—*la ville basse*—at the foot of the cliff, we can still spot the entrances to the limestone quarries. Many of the galleries were transformed into wine cellars, including the famous Cave Paincte ("Painted Cellar") that was already a tavern in the 1500s and is mentioned in the works of Chinon's prodigal son, François Rabelais (1494–1553). In his first novel, *Pantagruel* (1532), Rabelais has his ever-thirsty hero rave about the place: "I know where Chinon lies, and the painted cellar also, having myself drunk there many a glass of cool wine..."

Writer, physician, humanist, and freethinker, Rabelais is one of the first ambassadors of Chinon wine and of wine in general, "the most civilized thing in the world." This comes as no surprise, since he was born on the Vienne's left bank, where his family owned vineyards. The Rabelais also owned vines on the right bank, on the south-facing hill behind the Chinon castle. The place is now known as Le Clos de l'Écho ("Echo Vineyard"), in reference to the shouts of admiration bouncing off the walls as visitors and tourists walk up to the castle.

Rabelais' Clos de l'Écho now belongs to the Couly-Dutheil family. The estate produces a powerful Chinon, taking advantage of the limestone terroir, a well-drained, steep slope, and a southerly exposure. Garnet-colored with purple streaks, the wine has a cherry and black-currant bouquet, and well-structured tannins.

Indeed, the same terroir-ruled duality applies to Chinon as it does to Bourgueil: light and fruity "gravel wines" from the alluvial terraces, and robust "*tuffeau* wines" from the limestone hillslopes, aged in oak barrels. Since Bourgueil and Chinon share the same dual terroir, the same grape variety, and roughly the same climate, they are hard to tell apart. In fact, it is easier to distinguish a gravel Chinon from a limestone Chinon than a Chinon from a Bourgueil if they both come from the same type of soil—a perfect illustration of the role played by terroir in cultivating the character of a wine.

History doesn't say which Chinon type François Rabelais enjoyed the most, but there is little doubt that he drank loads of it, judging from the boisterous delirium running through his novels. According to him, "Juice from the vine clears the spirit and brings better understanding, appeases anger, banishes sadness and brings joy and merriness."

CHALK AND WINE

If we drive up the Vienne valley past Chinon, a road follows the flat, alluvial plain and its gravel bars. Parallel to it but higher up, route D 21 digs into the limestone slope and passes through a string of winemaking villages.

Cravant-les-Coteaux is one such village, 10 kilometers (6 miles) upstream of Chinon (fig. 7.3). On the outskirts of town, the road embankment offers a glimpse of the white, chalky limestone in cross section directly under the vines, showing how little soil the roots need to punch through before reaching bedrock.

Cravant-les-Coteaux is home to Bernard Baudry's estate, where father Bernard and son Christophe cultivate Cabernet Franc with a passion and make terroir a priority. Their vine lots are distributed on both gravel terraces and limestone *tuffeau*, and they vinify them separately. In the Baudry wine-tasting room, glass cases display a cross section of soil for each of the various climates they bottle. Here we clearly see and taste the contribution of gravel at Les Granges and Les Grézeaux: the former is light and fruity, to be drunk young, while the latter is slightly denser, since the vine stock is older and its roots penetrate deeper in gravel and clay, with the wine spending a bit more time in oak barrels before being bottled.

A significant change in character occurs when we move up the ladder—and the hillslope—to *tuffeau* wines, starting with La Croix Boissée from the limestone benches above the village. Bernard Baudry points out the wine's "chalky" bite: a puckering sensation that awakens our taste buds and prepares the palate for a powerful surge of red berries. This

Figure 7.3 Road cut at Cravant-les-Coteaux, upstream of Chinon. Under a few tens of centimeters of dark soil, the vines directly tap layers of white *tuffeau* limestone.

tuffeau wine will age well and over time develop notes of stewed fruit, spices, and forest undergrowth, depending on the vintage.

Different yet is Le Clos Guillot, a climate on the outskirts of Chinon, high up on the slope. The vines derive their aromas from a layer of yellow *tuffeau*, above the main benches of white *tuffeau*, with a top coating of sand and flint. The Baudry *tuffeau* wines thus bracket several million years of sea-bottom action as the sea came and went across Touraine. But before diving 100 million years back in time, let us take a few steps down into the Baudry cellar, where the last few years of production lie maturing in oak barrels lined up against the cool limestone walls. Here, each vintage has a story to tell, with the global climate sometimes overruling the subtleties of terroir.

To illustrate the point, Bernard Baudry has us taste a rare Chinon white, made from Chenin Blanc and dated 2004 (six years old at the time of our tasting). The wine gives off a surprising yet delicious earthy bouquet of asparagus and potato, which you can pick out in many other wines from that year, regardless of grape variety and terroir. There is a mysterious contribution of climate here: the soil was "baked" by an exceptional heat wave in the summer of 2003 and by a shorter one in June of 2004, perhaps a significant factor in such an unusual aromatic twist. This is not the only year when the weather overprinted French wines on a global scale: the 1985 vintage, for instance, was famous for a countrywide overprint of bell pepper.

THE CRETACEOUS SEA

Vintage notwithstanding, the Cabernet Franc of Chinon and Bourgueil celebrates gravel and limestone—a form of limestone that is the specialty of the Cretaceous period (145 million to 65 million years ago), when dinosaurs, flying reptiles, and swimming reptiles reached a peak in diversity, the separate evolutionary branch of birds developed, and small mammals proliferated. Within the Cretaceous, the limestone terroir of the Loire valley belongs to the upper part of the period, and precisely to the Turonian stage (93 million to 90 million years ago), named for the city of Tours and the Touraine region where it is so remarkably displayed.

The terroir of Chinon, with its layers of white and yellow *tuffeau*, gives us a picture of what France was like in the Upper Cretaceous (fig. 7.4). The climate was warm and tropical, with a new ocean basin opening up to the west: the Central Atlantic. The bulging of underwater volcanic rifts caused a new rise in sea level, further isolating French massifs from one another, separating them with shallow seas and lagoons.

The plankton calcareous ooze that settled on the sea bottom was spiked, as we saw, with mica flakes and opal beads injected by rivers draining the mountainous islands—the recipe for white *tuffeau*. Under much of the vineyard of Bourgueil and Chinon, its thickness reaches 40 meters (approx. 130 feet). Spanning a couple million years, the sedimentation occurred regularly but was by no means an avalanche: little more than 2 millimeters (a tenth of an inch) per century.

At the close of the Turonian age, the sea began to withdraw to the north, becoming much shallower in the Chinon area. A coarser, coastal limestone draped over the white *tuffeau*, with crushed seashells and more sand mixed in. Only a few meters thick, this upper *tuffeau* is yellow and crisscrossed by oblique bedforms indicating strong currents, and occasional hardpans that signal the temporary emergence of the seabed from the water. This yellow *tuffeau* underlies the highest-level vineyards around Chinon.

Figure 7.4 France during the Cretaceous period, 100 million years ago. The future site of the Loire valley vineyards lies underwater, building up layers of *tuffeau*: limestone mixed with minerals that eroded off the slopes of the Massif Central. (Map by Pierre-Emmanuel Paulis.)

The many fossils of this coastal *tuffeau* provide us with a good deal of detail about marine life-forms at the time. Besides oysters, we find scallops, sea urchins, and well-preserved ammonites. Although the ammonites are of modest size, there is an evolutionary trend toward larger species at the end of the Cretaceous (one ammonite species found in Germany is over 2 meters [6 feet] in diameter). On land, dinosaurs gave rise to large quadrupeds—the titanosaurs—while at the same time turning out a whole range of smaller bipedal species. The drifting apart of continental blocks during the Cretaceous also isolated animal groups from one another, and favored the development of new species from separate genetic pools. This was also true on a smaller scale in France, where the territory was cut up into a number of islands.

The French age of lagoons and tropical islands came to an end with an accelerated sea-level drop at the end of the Cretaceous, leaving in its wake sandy layers mixed with flint. This terminal deposit is found high up on the plateau, near the tree line. It constitutes a third, albeit minor type of terroir at Bourgueil and Chinon, as well as farther downstream on the site of another outstanding appellation of Loire valley wines: Saumur-Champigny.

SAUMUR-CHAMPIGNY

Also made from Cabernet Franc, Saumur-Champigny belongs to the same family of wines as Bourgueil, Saint-Nicolas-de-Bourgueil, and Chinon, but it is technically outside the Touraine region. It is centered on the riverside city of Saumur, which marks the beginning of the lower Loire region of Anjou.

SAUMUR-CHAMPIGNY AOC

Region:	Loire valley (Touraine/Anjou)
Wine type:	red, rosé
Grape variety:	Cabernet Franc
Area of vineyard:	1,500 hectares (3,750 acres)
Production:	85,000 hectoliters/year (1,130,000 bottles/year)
Crus:	none, but reputed "climates"
Nature of soil:	calcareous, sandy
Nature of bedrock:	chalky limestone (*tuffeau*)
Age of bedrock:	Cretaceous, Turonian stage (90 million years ago)
Aging potential:	2 to 10 years
Serving temperature:	14–18°C (57–64°F)
To be served with:	lamb, duck, game, red meat

Only 30 kilometers (less than 20 miles; a forty-minute drive) separate Saumur from Chinon, and their vineyards share the same grape variety, overall climate, and terroir. The likeness of the wines is therefore no surprise, but Saumur-Champigny has managed to build its own identity, topping the Cabernet's bouquet of raspberry and black currant with a distinct touch of violet.

The vineyard of Saumur-Champigny begins downstream of the picturesque village of Candes-Saint-Martin, on the left bank of the Loire valley, and stretches from Montsoreau westward to Saumur over a distance of 10 kilometers (6 miles). At Saumur, a small confluent flows in from the south—the Thouet River—which the vineyard also follows inland. The

Figure 7.5 Domaine de la Perruche, at Saumur-Champigny, is located on top a *tuffeau* limestone cliff riddled with "troglodyte" dwellings and lined with limestone houses, as represented on its label. (Photograph courtesy of Domaine de la Perruche, Montsoreau.)

third border of the triangular Saumur-Champigny terroir is the Fonte-vraud tree line.*

The vineyard's most scenic border is the Loire valley drive along route D 947 — "The Road of the Dukes of Anjou" — between Montsoreau and Saumur. Not only does it run along the lowermost vines planted on river gravel, but it offers a spectacular view of the towering limestone cliffs and troglodyte dwellings dug into the rock face (fig. 7.5). As mentioned earlier in the chapter, pueblo-type cliff dwellings were a way of life in the central Loire valley until very recently.

Upon reaching the village of Turquant (coming from Chinon), the most spectacular troglodyte galleries were living quarters for the nobility during the Middle Ages, and legend has it that Queen Margaret herself stayed here. Married at age sixteen to Henry VI of England, Margaret of Anjou (1430–1482) was involved in the wars of succession and political plots of the English Crown (the Wars of the Roses) as one of the main leaders of the House of Lancaster. After the murders of her husband and

*Nearby, you can visit the beautifully restored twelfth-century Fontevraud Abbey, complete with Romanesque church and cloister, where Eleanor of Aquitaine retired as a nun and was buried alongside her husband, Henry II of England.

her son, she fled to Anjou. There the queen spent the last years of her tragic life between Angers and Saumur, whose troglodytic dwellings and secret passageways offered a safe haven for her and her court.

Directly downstream of Turquant at Souzay-Champigny, there are more houses, cellars, and abandoned quarries carved into the rock face, including a quaint nineteenth-century miniature castle with turrets protruding from the cliff, and a signposted trail leading through abandoned galleries and dwellings. When you walk this underground trail, now silent and empty, you need to imagine it noisy and packed with people: a medieval mall with shopkeepers loudly hawking their produce and other foodstuffs, hammering blacksmiths, and mule-drawn carriages inching their way through the crowd. Fish and meat, fruit and vegetables, and barrels of wine benefited from the perfect storage conditions in the cool limestone cellars.

This underground lifestyle lasted until the late 1800s, when the railroad and modern life came to Saumur and the population began to desert the countryside and flock to the cities. Today, a small underground culture occupies the best troglodyte emplacements. As you walk the outside streets of Souzay along the cliff face, you will view sculpted façades of white *tuffeau*, vaulted doorways, cellars, and courtyards; on occasion a gate will lift, revealing a hidden driveway and swallowing a car into the rock face.

Some of the quarries and cellars have been converted into mushroom farms. The chalky substrate, low light level, and cool humidity are ideal for growing button mushrooms and shiitake, and the Touraine region was long France's leading mushroom producer. Now facing the competition of industrial greenhouses, *tuffeau* mushroom farms are fast disappearing—but if you pass through the Saumur district, do visit one of these fascinating underground establishments. Buy your mushrooms, order mushroom caps cooked in garlic at the grill, and of course accompany them with a glass of Saumur white or a red Saumur-Champigny.

Besides hosting food stalls and later serving the mushroom industry, the *tuffeau* quarries and galleries have been instrumental in winemaking. This underground world interfaces directly with the vine-covered surface via sloping ramps and vertical shafts. Walking through a vineyard on the plateau, it is not uncommon to run across a chimney top or a TV antenna sticking out of the ground (conversely, you can imagine living in an underground dwelling and noticing vine roots poking through the ceiling). This double-sided exploitation of *tuffeau* limestone, from above and below, provided vine growers with clever shortcuts. Harvesters could dump the grapes directly down a chute to the underground city below, and the wine was pressed, stored, and sold on-site in the troglodyte market: vertical integration at its best.

THE LIFE AND WORKS OF ANTOINE CRISTAL

The rise to fame of Saumur-Champigny wine can be credited to one man in particular: Antoine Cristal (1837–1931). Born in the village of Turquant, Cristal began his career as a traveling salesman, investing his earnings in the stock market and amassing a small fortune. This wealth allowed him to buy the Château of Parnay between Turquant and Souzay-Champigny, and at age fifty start a new life as a winemaker.

At the time, the Saumur area produced mostly white and sparkling wines. Sparkling wines, made from Chenin Blanc, are still around under the label Saumur Mousseux, and are a great alternative to Champagne at one-third of the price (around 5 euros a bottle). But it is Antoine Cristal who put red Saumur on the map. The phylloxera blight devastated European vineyards in the late 1800s, and Cristal was among the first to replace the disease-ridden vines with American rootstock, naturally resistant to the larvae. He seized the opportunity to experiment by grafting Cabernet Franc onto this rootstock, to see if he could compete with Bourgueil and Chinon, his established upstream neighbors. Moreover, he took a stand against artificial methods to improve the wine with added sugar, arguing that "our hillsides, with little humus and porous *tuffeau* bedrock, are built to give us a natural wine, and the sun alone suffices for that!"

On the other hand, Antoine Cristal had a curious mind and loved to tinker in his workshop, inventing new tools and devices for tending to his vines. He was one of the first winemakers to stretch iron wires across his vineyard to train the vine branches horizontally. But his most ingenious invention is a wall system he first erected as an experiment on half a hectare (roughly one acre) of vines, which he named Le Clos d'Entre les Murs ("The Walled-in Lot").

There was nothing new in erecting stone walls along a vineyard, either to claim ownership of a lot or to reflect sunlight onto the closest vines. But Cristal took the concept a step further by having his masons build eleven parallel, east–west walls within the enclosure, and by planting the vines on the northern side of each wall, in the shade. Half a meter aboveground, a hole was made in every parallel wall and the vine trained through it to emerge on the southern side, where its leaves and grapes would be in full sunlight. In this fashion, the vines, according to Cristal, kept "their feet cool and their belly warm."

Akin to a solar oven, the walls heated the grapes and brought them to maturity much faster than those raised in the open fields—their harvest occurred up to a month sooner. No wonder Cristal needed no sugar to spike his wine!

Happy with the outcome of this experiment, the winemaker had a second, larger lot built with parallel walls, although these were more widely spaced than in the first wall system. This he had built on the limestone slope right outside the tiny village of Champigny, which he bequeathed to the Hospices of Saumur in 1929, two years before his death.

In the meantime, Cristal and his wines rose to stardom, and a number of his bottles ended up in the cellars and on the tables of European aristocrats, including the tsar of Russia and the prince of Wales. French senator and prime minister Georges Clemenceau (1841–1929) met the outspoken winemaker during a trip through the Loire valley and later wrote to him: "Your vineyards are infinitely curious and the outcome of a powerful methodology, but even more interesting to me is the personality of their creator. . . . This is why I will not forget my visit to your château of Parnay."

Upon Cristal's death, the château and its vineyard slowly sank into oblivion. They were recently purchased by two passionate winemakers—Mathias Levron and Régis Vincenot—who now produce excellent Saumur-Champigny wines on the estate, including a well-balanced and moderately priced Clos du Château from vines directly outside the buildings, on the slope leading up to the plateau. Next up the scale is a Cuvée Historique, more expensive but truly remarkable, and finally, the famous Chenin Blanc white Saumur, grown within the walled acre of Clos d'Entre les Murs: a collector's item, since only two thousand bottles of it are produced each year.

After tasting the wine (and, no doubt, filling the car trunk), you can visit the galleries leading off the château courtyard into the cliff face, where Antoine Cristal's hundred-year old winepress stands. Next, we drive up the hill to the Clos d'Entre les Murs, and peek into the walled vineyard.

High above the Loire valley, we catch a nice view of the village church spires rising above the vines. Turning to the south, we'll see more vines running inland to the wooded horizon, and the little side road leading to the hidden village of Champigny that gave its name to the appellation. In the summertime, the sun beats down on the plateau, creating a fiery microclimate. In fact, the very name of Champigny allegedly derives from the Latin *campus ignis*: "fields of fire." If we are courageous enough to hike the vineyards, we might well finish our trek in the coolness of a *tuffeau* cellar, enjoying a well-deserved glass of Saumur-Champigny.

CHAPTER EIGHT

Vineyards of Provence

In Provence, the vineyards encompass a broad geographical area, from the mouth of the Rhône River—the Camargue delta—eastward to the Italian border. Their geological mosaic of terroirs derives from rock strata of all types and ages, and they feature a wide array of grape varieties that sound and taste Mediterranean: Grenache, Syrah, Mourvèdre, Tibouren, Carignan . . .

It might seem difficult at first to find a unifying thread running through such a variety of grapes and places, but several factors do come together to bring an unmistakable cachet to the wines of Provence: the wines' distinctive grape varieties; the Mediterranean climate that blends summer heat, mistral wind, and maritime influence; and the ancestral knowledge of the winemakers themselves.

The first evidence of winemaking in the region dates to the sixth century BC and is attributed to Phoenician merchants and the Phocaeans—Greek sailors from Anatolia (modern-day Turkey) who founded the city of Marseille in 600 BC, according to legend. Scores of amphorae discovered on the seafloor attest to this early wine trade. Throughout the centuries, Greek civilization and commerce spread the art of winemaking along the Provence coastline; and when the Romans replaced the Greeks, they found a well-established infrastructure, complete with trade routes penetrating deep into France—one of the main reasons for Julius Caesar's conquest of Gaul, according to some historians.

Today a booming business, the wines of Provence are best known for their light rosé brands that fill the aisles of French supermarkets in the

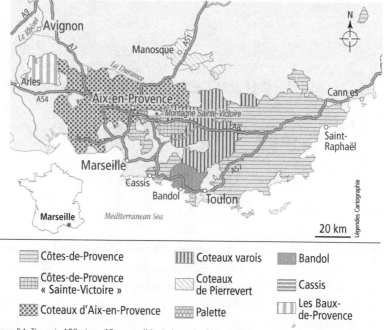

Figure 8.1 The main AOC wines of Provence. (Map by Legendes Cartographie.)

summertime, but that's only part of the picture (fig. 8.1). The region also produces sturdier, top-notch rosés; superb dry whites, such as the appellation Cassis; and strong aromatic reds, such as Bandol, Palette, and a number of interesting Côtes-de-Provence, including the newly recognized Côtes-de-Provence Sainte-Victoire.

THE GEOLOGICAL SAGA OF PROVENCE

The land that is now Provence first began building its rocky basement over 500 million years ago, through the rise of great blobs of granitic magma that infiltrated old fracture zones and pushed to the surface to erupt in lava flows and clouds of ash. This output was compressed and transformed into layers of dark mica-schist. One good example is the Maures Massif between Toulon and Fréjus, whose cores of granite and layers of schist are covered with scrubland (locally named *garrigue*) and forests of pine and oak. Winemakers have planted along its seaside as well as inland, producing mostly fresh and light rosés, such as those from the Saint-Tropez peninsula.

Cinsault is the dominant grape variety here, allegedly born in Provence; it is well adapted to arid conditions and to poorly developed dry soil. A black grape, it is often blended with other varieties in producing Provence reds, but above all it is the leading varietal of Provence rosés.

But let us continue with our geological overview of the area. Around 300 million years ago, the region underwent extension and rifting, resulting in a spectacular surge of volcanic activity. This additional pileup of lava and ash—the Esterel Massif—stretches east of the Maures Massif from Fréjus to Cannes. The ash layers, turned golden orange, have been carved by erosion into spectacular slopes and cliffs tumbling down to the blue Mediterranean Sea, best viewed from the coastal railroad or from the winding RD 98 motorway.

Few vines are in this volcanic range—the Coteaux Varois appellation ends short of Fréjus—but the oak trees that grow readily on this alkaline, silica-rich soil are still harvested for their thick, porous bark, which is used in the manufacture of wine corks. Vineless as it is, the volcanic Esterel still makes an essential contribution to wine.

Only later in our geological sequence, during the Triassic period (250 million to 200 million years ago), did Provence's terroir build up substantially, as marine inlets penetrated the old volcanic ranges and spread into shallow, saline basins, precipitating gypsum and other salts, collecting sand brought down by the streams, and piling up layers of limestone made of plankton tests and seashells that proliferated in the warm tropical waters. The Alpine mountain range did not yet exist, and a great marine gulf occupied eastern Provence—an offshoot of the Tethys Sea, precursor of the much later Mediterranean Sea.

The bedrock and terroir of Provence were completed undersea during the Jurassic and Cretaceous periods, 200 million to 100 million years ago. The limestone and marl strata, now uplifted, jut out as high cliffs along the seashore, indented by dry riverbeds and fjord-like *calanques* east of Marseille, and as great folds forming isolated inland mountain blocks, such as the Sainte-Baume Massif and Montagne Sainte-Victoire. This limestone terrain is home to two of the best wines of Provence: Cassis and Bandol.

BANDOL

The vineyards of Bandol are among the oldest ones in France, allegedly planted in the sixth century BC by Phocaean Greek merchants on a natural amphitheater overlooking the Mediterranean between La Ciotat and Toulon—with the Bay of Bandol providing an outlet for the wine trade. When the Romans took over the vineyards, they stabilized the gully-ridden steep slopes by way of walled terraces, known in Provence as *restanques*. These structures keep in check the erosion of a terroir made of sandstone, marl, and limestone from the Late Cretaceous (90 million to 85 million years ago), with a little gypsum and red clay thrown in.

The wines of Bandol owe their quality to the terroir, to an exceptional climate—abundant sunlight, marine coolness, and the drying effect of the mistral wind—and to an original grape variety, Mourvèdre, that takes full advantage of the conditions. Rather capricious, in need of sun but also of the cool sea breeze that slows down its ripening, Mourvèdre reaches its very best at Bandol. In winemaking it is touched up with a bit of Grenache and Cinsault, which contribute elegance and suppleness to the blend.

Bandol is a pleasant, strong red wine when young, with a palette of aromas ranging from cherry to black currant, pepper and violet. It also ages well—up to ten years—and develops aromas of licorice, tobacco, leather, and forest undergrowth. Charged with tannin, it travels well, which spread its reputation far and wide: it was one of the favorite wines of French king Louis XV (1710–1774). Red wines are the Bandol region's claim to fame, but represent only 30 percent of the production; the market favors rosé (65 percent), based on the same grape varieties and boasting a fresh, peppery flavor. Rare Bandol white (5 percent) is made from Ugni Blanc, Clairette, and Bourboulenc, grown on the northern slopes of the vineyard (making do with a lesser amount of sunlight) and developing a bouquet of white flowers and citrus fruit.

THE VINEYARDS OF CASSIS

Whereas Bandol is famous for its red and rosé wines (see box), Cassis is the home of great whites. Its vineyard (fig. 8.2) encompasses only 200 hectares (500 acres)—smaller than New York's Central Park—that form a horseshoe around the coastal village, with its two wings climbing the limestone slopes on either side of a wooded crest (Les Rampides).

The Cassis vineyard was probably founded by the Phoenicians or the Phocaeans, but it was during the Renaissance that it rose to fame under the stewardship of the Florentine Albizzi family. It was known for its reds as well as for its dry and sweet whites, pressed from Muscat grapes. One of the twelve or so domaines of Cassis bears the name of the illustrious family: Le Clos d'Albizzi.

Cassis could well have thrived on its diverse production, had it not been for the phylloxera blight that destroyed the vineyard in the 1890s.

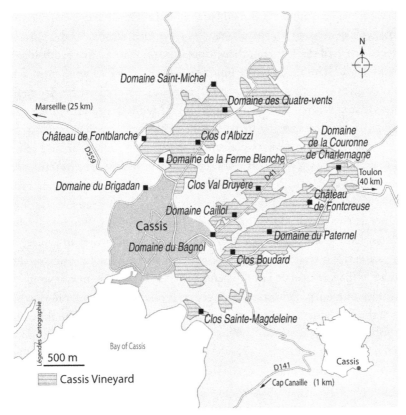

N

Domaine Saint-Michel

Marseille (25 km)

Domaine des Quatre-vents

Château de Fontblanche

Clos d'Albizzi

Domaine
de la Couronne
de Charlemagne

Domaine de la Ferme Blanche

Domaine du Brigadan

Clos Val Bruyère

Toulon
(40 km)

Château
de Fontcreuse

Domaine Caillol

Cassis

Domaine du Paternel

Domaine du Bagnol

Clos Boudard

Clos Sainte-Magdeleine

Bay of Cassis

500 m

Cassis

Légendes Cartographie

Cassis Vineyard

D141

Cap Canaille (1 km)

Figure 8.2 The Cassis AOC vineyard and its fourteen estates. (Map by Legendes Cartographie; modified from the map of Syndicat des vignerons de Cassis.)

When new rootstock resistant to the bug was imported from America and planted at Cassis, it was discovered that Muscat did not graft well onto it, whereas Marsanne and Clairette fared much better, steering the production away from sweet to dry whites that were highly praised. The quality of the wine, paired with the long history of the vineyard, made Cassis one of the first wines ever to receive the Appellation of Controlled Origin (AOC), sharing the first short list of 1936 with Arbois, Sauternes, and Châteauneuf-du-Pape.

The specificity of Cassis holds in its ancestral winemaking tradition, limestone- and marl-terraced terroir, and remarkable microclimate with abundant sunshine, moderated by a cool sea breeze that slows down the ripening of the grapes and lends complexity to the wine.

If we drive to Cassis from Marseille along the D 559, one of the first estates we will encounter at the side of the road is the Domaine de la Ferme Blanche. Seated up on the plateau, its rows of Marsanne and Clairette recede toward the horizon, their roots firmly anchored in a reddish soil stuffed with chunks of white limestone. A classic white wine from this estate gives off typical scents of eucalyptus and menthol, with a dry coolness that makes it a perfect match for local Mediterranean fish, such as a *loup* (sea bass) or a *sar* (white sea bream) grilled in olive oil.

Rosé wines represent 25 percent of the Cassis production, and at La Ferme Blanche they give off a floral bouquet of fruit tree blossoms, rose, and honeysuckle, with a touch of lychee. Cassis rosé complements Provençal, Moroccan, or Indian cuisine very nicely. As for red Cassis (only 2 percent of the production), it relies, like its neighbor Bandol, on the Mourvèdre grape variety and breathes out aromas of leather and spice.

After this introduction to Cassis wines, it is worth backtracking up the road to the Château de Fontblanche, protected from the mistral wind by a limestone scarp. The estate belonged to Émile Bodin (1881–1969), who was the first winemaker in Cassis to plant American rootstock in the early 1900s, in the wake of the phylloxera blight. Provence's celebrated poet Frédéric Mistral (1830–1914) raved about his friend's wine: "It shines like pure diamond and smells like the rosemary, myrtle and heather that cover our hills and dance in our glass."

On this arid terroir where the rock breaks down into a reddish soil of limestone, sand and clay, the Château de Fontblanche produces dry, fruity whites, with Marsanne bringing richness and good length on the palate, Clairette and Ugni Blanc a touch of acidity and freshness, and Bourboulenc a velvety texture. Depending on the proportions, the aromas vary from smoky flint to white flowers, almond, and honey. Reds and rosés rely on Mourvèdre, Grenache, and Cinsault to bring out aromas of raspberry and spices for the red, strawberry and almond for the rosé.

This sampling of Cassis vineyards would not be complete without a stop at Le Clos Sainte-Magdeleine, which enjoys a spectacular location on the seafront and a reputation for excellence. Its elegant Provençal mansion, surrounded by vines and pine trees, is set on a spit of limestone between two coves of turquoise water, at the foot of the mighty Soubeyrannes cliffs. In this grandiose setting, Le Clos Sainte-Magdeleine produces a white Cassis that ranks among the best white wines in France and is kept in stock in the cellar of the Élysée presidential palace in Paris.

Treated to a round of wine tasting on the estate, we learn from Anne-Marie Beranger, the Master Blender's wife, that the white is a blend of 50 percent Marsanne, 25 percent Ugni Blanc, and 25 percent Clairette, pressed separately. Pale yellow with greenish streaks, it gives off a fruity and mineral bouquet with a touch of white pepper, and flatters the palate with mineral freshness and a velvety finale of honey and pineapple. Sainte-Magdeleine's rosé favors Grenache (40 percent), Cinsault (40 percent), and Mourvèdre (20 percent), with a salmon-pink color and a bouquet of red berries.

One-fourth of the vines, for both the white and the rosé, are planted in the closed lot surrounding the mansion on the seafront spit, and the balance on higher terraces up the slope of the Soubeyrannes cliff. On the spit, not much soil covers the limestone bedrock, so vines must extend their roots deep into crevices, seeking the freshwater that rises through capillary action during the hot summer months.

The wines of Cassis owe a good deal of their personality to these calcareous rocks underpinning the vineyards. They were laid underwater some 100 million years ago in the Middle Cretaceous period, when a warm and shallow sea covered the south of France from the Atlantic eastward to the Tethys Sea—the ancestor of the Mediterranean—extending its northernmost gulf across Provence and up into a basin that would later give birth to the Alps.

PROVENCE EMERGES FROM THE SEA

Provence's shallow sea was bounded to the north by shoals that ultimately rose out of the waters—known to geologists as the Durance bulge—and to the south by another landmass consisting of Corsica and Sardinia, tucked alongside Provence at the time (before splitting away from the mainland tens of millions of years later).

During this phase of the Cretaceous, known as the Cenomanian stage (100 million to 94 million years ago), the shallow waters on the site of Cassis received a substantial input of sand and clay from nearby Corsica, but it was mostly the marine sea life that dominated the sedimentary regime. Bivalve seashells packed together in dense colonies known as rudists built massive reefs. Today, they stand out in the landscape as hard and steep limestone ledges, jutting out from softer strata above and below.

At Clos Sainte-Magdeleine, when you look up at the Soubeyrannes cliffs, you get a sense of the immense buildup of the rudist reef over millions of years. Breaks in the slope, showing a shallower profile, indicate periods when sedimentation switched to softer material, now prone to slumping. These shallow benches of blue marl form the terraces hosting the highest Sainte-Magdeleine vineyards.

It so happens that the marly terraces mark a turning point in the Cretaceous calendar: the Cenomanian-Turonian boundary, dated at 94 million years before present. The boundary corresponds to a noticeable shake-up of the worldwide ecosystem: nearly 25 percent of all marine species died out at this level, including the last genus of dolphin-shaped ichthyosaurs. The crisis is attributed to a high stand in sea level, which connected shallow habitats and decreased species diversity through competition. Perhaps underwater volcanism also triggered a water-warming greenhouse effect, and drinking the wine from the boundary might give geologists some other ideas as well.

Above the sloping marl terroir, the rest of the cliff face is vertical up to the top, and laced with benches of red, pink, and white limestone. If we drive up the road and follow the edge of the cliff eastward, we reach Cap Canaille, towering 400 meters (1,300 feet) above the Mediterranean—the highest cliffs in France. From this vantage point, we have a great view of the eastward extension of the Soubeyrannes cliffs, sloping down toward the Bay of La Ciotat and its monumental shipyards. If you look inland to the north, you can make out on the horizon the folded crest line of the Sainte-Baume Massif, one of the many mountainous wrinkles that run east to west across Provence. These folds are also made of limestone and sandstone that were deposited underwater and later uplifted by tectonic forces that affected Provence at the end of the Cretaceous—a compression vector oriented south to north, although its origin can be traced westward to the expansion of the Atlantic Ocean.

Far to the west, the Atlantic basin had indeed been expanding throughout the Jurassic and the Cretaceous, rotating Africa away from North America. Since Spain was attached to Africa, it was pulled away from France, rotating counterclockwise and opening up the Gulf of Biscay between them. By the end of the Cretaceous, the swinging motion of Spain ended up compressing the land to the east, putting southern France under pressure and creating the Provence mountain ranges.

THE SAINTE-BAUME MASSIF

From Cassis and La Ciotat, highway A 50 takes us 15 kilometers (9 miles) north to the city of Aubagne and the Sainte-Baume Massif. This spectacular limestone and sandstone mountain ridge stretches 20 kilometers (12 miles) east to west and rises 1,147 meters (3,823 feet) above sea level. Its southern flank falls steeply on a plateau of dry scrubland, whereas the shallower slopes of its northern flank are covered by a primordial forest of Scots pine, lime, maple, and beech trees.

The Sainte-Baume Massif helped early geologists uncover the laws of structural geology. In 1865, while mapping the mountain, geologist Henri Coquand established that its upper layers were old Jurassic limestone that rested on younger Cretaceous rock, contrary to the logical rule that sedimentary layers are always older at the bottom and younger on top. What could possibly explain such an inverted sequence? One had to assume, which Marcel Briand boldly did in 1890, that the thick pile of Jurassic rocks had climbed onto the back of younger Cretaceous strata through some unknown mechanism.

We now know this to be the case: the rotation of the Iberian block compressed, lifted, and detached huge slices of rock from their basement, causing them to slide and creep dozens of kilometers northward across Provence and cover younger sediments in the process. Known as nappes, these gigantic overthrusts have now been recognized the world over in most mountain chains that have undergone tectonic compression, such as the Alps and the Himalayas. In Provence, they take the form of rock "waves" in the landscape, separated by plains and valleys.

The northern flank of the Sainte-Baume Massif slopes down to the valley of the Huveaune—a river passing through the towns of Sainte-Zacharie and Auriol—before another wave of limestone swells up to form the Regagnas and Auréliens Mountains. This rolling topography is well adapted to vine growing, especially those lower slopes that combine limestone substrate, a fair amount of sunshine, and access to water stored at depth in the permeable rock. The Sainte-Baume Massif acts as a water tower for the area, collecting rainfall on its upper slopes and funneling the water down through an underground karstic fracture system before finally discharging it at the foot of the slope in the form of resurgent springs.

Vines are planted all along the lower slopes of Sainte-Baume and westward into the hills of Garlaban, where local wine producers from the villages of Saint-Zacharie, Auriol, La Destrousse, and Gémenos joined forces to found a cooperative, Les Vignerons du Garlaban, and make the most of the terroir. Planting noble grape varieties such as Grenache and Syrah for their red and rosé, and Rolle for their white, they turn out rich, aromatic wines that are fully representative of Côtes-de-Provence as a whole.

SAINTE-VICTOIRE: THE MOUNTAIN

After climbing over Sainte-Baume, descending into the Huveaune valley, and rising over the Auréliens Mountains, our roller-coaster journey once again takes us down into a wide, filled-in valley—the Arc river basin—facing a new mountain front in the distance, which stretches 12 kilometers (7 miles) east to west across the horizon: the celebrated Montagne Sainte-Victoire (fig. 8.3).

Located east of Aix-en-Provence, Sainte-Victoire is world famous, thanks to postimpressionist artist Paul Cézanne (1839–1906), who immortalized its hat-shaped profile in nearly one hundred paintings (forty-four oils and forty-three watercolors) on exhibit around the globe. Cézanne considered the mountain a living creature, a moody sphinx that changed aspect from one moment to the next, depending on the cloud cover and lighting. He developed a scientific curiosity about it, as he confessed in a letter to his friend Joachim Gasquet: "I need to learn geology and understand how Sainte-Victoire is rooted."

As is the case for other mountain ridges of Provence, Sainte-Victoire is made of Jurassic and Cretaceous limestone and marl accumulated underwater before they were lifted above sea level around 100 million years ago. The east–west-trending high grounds separated a southern sea branch pushing in from the Atlantic to the west, and a northern sea branch coming in from the east—an extension of the proto-Mediterranean Tethys Sea.

Named the Durance Isthmus by geologists, the limestone peninsula that rose between the northern and southern seas became covered with tropical vegetation. Abundant rainfall washed out the land, creating a reddish iron and aluminum-rich soil similar to today's lateritic soil of South America and equatorial Africa. Starting in the early 1800s, pockets of this red soil were mined in Provence for the aluminum they contained, namely

Figure 8.3 At the foot of Montagne Sainte-Victoire, Provence vineyards are rooted in End Cretaceous clay and limestone.

around the medieval town of Baux-de-Provence—hence the name baux-ite for aluminum ore—and along the Durance Isthmus as far east as Montagne Sainte-Victoire.

In the Late Cretaceous, the site of Sainte-Victoire was not yet the hat-shaped mountain that we admire today, but it already showed substantial elevation: we know this because of layers of stream-carried pebbles that lie nearby, dating to the period and pointing to steep slopes as their source. There was also a marine gulf extending to the foot of the mountain, getting shallower with time as the sea level dropped at the close of the Cretaceous. This gulf gave way to a string of lakes and swamps that accumulated brackish and freshwater limestone and red clay.

Much later, during the Alpine upheaval, all these layers were uplifted and folded, giving shape to Montagne Sainte-Victoire as we now know it. These tectonic displacements fortunately spared most of the sediments involved, so that we can still read into the rock and decipher the closing events of the Cretaceous period in Provence.

This Late Cretaceous story is most important, since it records the passing through of many dinosaur species, up until the brutal mass extinction that wiped out the giant reptiles along with many other animal groups, paving the way for the rise of mammals and ultimately the ascent of man.

This change of era, from the reptile-dominated Mesozoic to the mammal-ruled Cenozoic, is of itself an excellent reason to explore Montagne Sainte-Victoire and its sedimentary record. But there is another good reason to hike the mountain: grapevines do very well in Late Cretaceous and Early Cenozoic red clays, delivering superb wines that have won their own exclusive AOC appellation of Côtes-de-Provence Sainte-Victoire.

SAINTE-VICTOIRE: THE VINEYARD

Côtes-de-Provence as a whole was awarded its official regional AOC appellation in 1977, but over the years local winemakers rightly claimed that the wide area also includes specific, high-quality terroirs deserving of their own restricted brand name. To apply for separate appellations, they filed technical reports in which they assembled information on the historical track record and savoir-faire of their candidate locations and the microclimate present at each, along with a list of recommended grape varieties. One of the terroirs that emerged from the lot was the new appellation Côtes-de-Provence Sainte-Victoire, officially recognized in 2005 and applicable to red wines (20 percent of the production) and rosé wines (80 percent). This terroir involves nine townships on the southern flank of the Sainte-Victoire Mountain and over 2,000 hectares (5,000 acres) of vineyard owned by twenty-two estates and winemaking cooperatives.

Whereas the regional Côtes-de-Provence AOC limits production to 55 hectoliters per hectare (3.57 tons per acre) and allows the use of Grenache, Syrah, Cinsault, Mourvèdre, and Tibouren grapes (with possible minor additions of Cabernet-Sauvignon and Carignan), the more select Côtes-de-Provence Sainte-Victoire favors quality over quantity, restricting productivity to less than 50 hectoliters per hectare (3.25 tons per acre) and limiting the range of grape varieties to mainly Grenache, Syrah, and Cinsault.

Côte-de-Provence Sainte-Victoire also boasts a microclimate that affects the southern slopes of the mountain down to the Arc River, as we learn by visiting one of the key estates of the appellation: Mas de Cadenet.

Right off road D 57, between Trets and Puyloubier, Le Mas de Cadenet owes its name to the juniper tree—*cade* in Provençal—a typical shrub of

Provence. *Mas de cadenet* means an estate planted with juniper, but that was before vines took over and replaced the scrubland.

Le Mas de Cadenet was instrumental in putting Côtes-de-Provence Sainte-Victoire on the map and into the wine books, since it is the Negrel family, owner of the estate, who worked hardest to file the request for the appellation and get it passed. Matthieu Negrel once took me on a tour of his vineyard and pointed out the importance of the microclimate, starting with the luminous wall of limestone raised by Sainte-Victoire to the north. The obstacle takes the edge off the cold mistral wind, but lets enough of it through to blow-dry the vines and ward off any fungi and diseases linked to humidity.

Playing a similar role on the southern horizon, the Auréliens mountain range, which we crossed on our way from Sainte-Baume, cuts off the sea breeze blowing in from the south, so that rainfall in the vineyard is kept to a minimum (fig. 8.4). Even summer storms spare Le Mas de Cadenet, which is far enough from the high slopes and their zones of turbulence. Thunderstorms gather on the eastern flank of the mountain and spare the vines in the lowlands.

Le Mas de Cadenet produces red, rosé, and white wines. Although only red and rosé comprise the Sainte-Victoire label, the white is none-

Figure 8.4 Looking south, the vineyard of Mas de Cadenet estate is protected from humid air and rainfall by the Auréliens Mountains on the horizon.

theless spectacular in its own right. Produced from Rolle grapes and aged in oak barrels, it gives off a bouquet of beeswax, bergamot orange, vanilla, pear, and citrus fruit.

The estate's rosé, labeled *Mas Negrel Cadenet*, is a blend of 40 percent Grenache, 40 percent Cinsault, and 20 percent Syrah, with a bouquet bringing out nutmeg, quince, and candied orange: a wine that is a great match for fish and shellfish, as well as lamb and duck. As for the Mas de Cadenet red, made from Grenache (45 percent), Syrah (45 percent), and a touch of Cabernet-Sauvignon (10 percent), its bouquet of black berries and spices makes it a natural for red meat, such as a barbecued beef rib.

Terroir-wise, you get a good idea of the soil texture right outside the Mas courtyard, lined with vines. The ground is littered with cobbles washed down from the distant Sainte-Victoire Mountain or rolled in long ago by high stands of the Arc River. The geological map tells us that this pebble cover is only superficial, and that underneath it we should hit red clay from the End Cretaceous, laid down in the swamps surrounding Sainte-Victoire at the time. Confirming this, Matthieu Negrel brings out a chunk of clay he found on his lot; it contains the protruding fossilized shell of a dinosaur egg.

CÔTEAUX-D'AIX-EN-PROVENCE

The vineyards of Côtes-de-Provence Sainte-Victoire occupy the southern flank of the Sainte-Victoire Mountain down to the Arc riverbed. The distinct appellation Côteaux-d'Aix-en-Provence starts on the western flank of the mountain and stretches as far west as the Berre inlet (Étang de Berre) on the Mediterranean coastline, and inland to Salon-de-Provence and the Alpilles mountain range. Two-thirds of the production is red wine and a third is rosé, with Grenache the dominant grape variety; minor contributions come from Cinsault, Carignan, Mourvèdre, and Cabernet-Sauvignon. The rosés are light and fruity, while the reds are more robust, with good aging potential.

The terroir of Côteaux-d'Aix is mainly limestone—as opposed to the dominant clay at Sainte-Victoire—folded into high hills along the same east–west axis as Sainte-Victoire and Sainte-Baume. One interesting exception is the Beaulieu terroir, nestled within the worn-down crater of a 20-million-year-old volcano; it features permeable basaltic scoria.

At the westernmost edge of Côteaux-d'Aix, the vines climbing up the slopes of the Alpilles range are entitled to the separate appellation Baux-de-Provence; they occupy a limestone terroir laced with veins and pockets of red ochre and aluminous bauxite. Here we are getting close to Côtes-du-Rhône, both geographically and wine-wise: the Baux-de-Provence reds are full bodied with a bouquet of plum and cherry.

Figure 8.5 The herbivorous, parrot-billed dinosaur *Rhabdodon* lived at the foot of the paleo-Sainte-Victoire mountain, 70 million years ago. (Photographed at the Museum of Natural History of Aix-en-Provence.)

VINES AND DINOSAUR EGGS

The first dinosaur fossils were discovered in the Marseille area in the 1840s. Provençal geologist Philippe Matheron (1807–1899) at the time identified two dinosaur types: a large, lizard-hipped titanosaur, later classified as part of the genus *Hypselosaurus*, and a bird-hipped and parrot-beaked ornithopod classified as *Rhabdodon* (fig. 8.5), along with a number of fossilized eggshell fragments.

Dinosaur bones are found mostly in riverbed sandstone, the carcasses having rolled along the river bottom and gotten covered up with sand, whereas the eggshells (fig. 8.6) are found in the fine red clay that lined the Cretaceous swamps—obviously a nesting favorite of female dinosaurs. Many of the eggs are clustered at the top of clay mounds that the animals built to place their eggs above flooding high stands; they covered the eggs with dirt and branches to protect them and speed up their incubation.

The dinosaur-egg clay corresponds to the last stage of the Cretaceous period, named the Rognacian in Provence because it was mapped around the town of Rognac above the Berre inlet. Worldwide, however, this stage is better known as the Maastrichtian, since the official "stratotype" was

Figure 8.6 The terroir of Côtes-de-Provence Sainte-Victoire is made up in places of End Cretaceous clay containing dinosaur egg fossils. (Photographed at the Museum of Natural History of Aix-en-Provence.)

defined at Maastricht in the Netherlands. It covers the time interval running from 70.5 million to 65.5 million years ago, at the end of which all the dinosaurs became extinct.

The dinosaur-egg clay outcrops all along the southern flank of Sainte-Victoire Mountain from Tholonet to Pourrières, splitting into two wings that encircle the younger, higher-standing Cengle Plateau, built of more resistant lacustrine limestone. The clay colors the soil red and is best viewed in cross section where gullies have cut through it, or where little mounds of it are left standing in the plains, spared by erosion.

Near the town of Rousset, a few of these buttes stand alongside the road, crowned by pine trees. Here in the 1950s, the first digs were carried out to collect dinosaur eggs. The main egg form is named *Megaloolithus mamillare*, and there is no telling which dinosaur species it belonged to, although there is a good chance it could be of the genus *Hypselosaurus* — the tall and long-necked titanosaur identified by Philippe Matheron that measured 12 meters (40 feet) long and weighed up to 10 tons.

Many vineyards are planted directly on the fossil-rich clay on either side of route D 7N. This is the case for Château de la Bégude on the northern side of the road in the township of Rousset, at the route's intersection with D 56C that heads north toward Sainte-Victoire.

A meeting place of the Knights Templar, then a coaching inn in the days of Louis XIV, the building and its vineyard came to be owned by a Catalan winemaker, Marius Rey, in the early 1900s. Now his great-grandson Jacques Lefebvre runs the estate. A former professor of English literature, Lefebvre sponsors winemaking workshops on his domaine and drives his own truck to sell his production across France, door to door; you will find his Château de la Bégude in a number of the country's restaurants.

There are dinosaur eggs in Jacques Lefebvre's vineyards, as in many other Rognacian outcrops around Sainte-Victoire. The winemaker has collected some fine specimens on his land over the years, but private property is no place for tourists to go fossil hunting. Here, it is the wine that you come looking for. If it's fossils that you seek, it's best to head north past the limestone Cengle Plateau, to the mountain-girdling route D 17.

Along this picturesque Route de Cézanne, as it's called, you will find a turnoff into a little parking lot servicing the Natural Park of Roque-Haute. A footpath takes you across a stream to a knoll of red clay planted with pines, where hikers stop to picnic and rest in the shade. If you look around closely, you will spot thin, pearly circles standing out against the red clay: these are dinosaur eggs, seen in cross section. Do not try to dislodge them from the hard ground, for you will only end up shattering the fragile remains. Try watering them instead: pouring water over the spot will darken the red clay and lighten the pearly shells, bringing out the shape of the eggs.

If you use up all your water playing with dinosaur eggs, one place to quench your thirst is the Domaine de Saint-Ser, 3 kilometers (2 miles) up the Cézanne Road toward Puyloubier. This is the highest vineyard on the mountain, where owner Jacqueline Guichot will have you taste her varied production of red, white, and rosé Côtes-de-Provence and Côtes-de-Provence Sainte-Victoire.

Here, too, at Saint-Ser, red clay and dinosaur eggs are underfoot, but they are hidden by a cover of older limestone blocks, shed off the mountain face. Wherever you look, the view is breathtaking. Uphill, Sainte-Victoire puts up a blinding white wall against an often moody sky; downhill to the south on a clear day, you can view the vineyards in the foreground, the Cengle Plateau at midslope, the Arc valley down in the flats, and the Auréliens Mountains on the horizon. Up at Saint-Ser, there is also a garden planted with the nineteen different grape varieties grown in Provence—an occasion to brush up on your varietals.

Grenache, Cinsault, and Syrah are the varieties retained for the estate's rosé, contributing respectively a bouquet of red berries, finesse, and aromatic intensity to the wine. Only Rolle is used for the white, giving off a bouquet of grapefruit and pineapple, with a woody touch of vanilla provided by a few months of repose in oak barrels. As for the Sainte-Victoire red, it marries Grenache with Syrah and Cabernet-Sauvignon, or only Syrah and Cabernet-Sauvignon for the Hauts de Saint-Ser cuvée. Syrah procures a powerful, spicy aroma, and Cabernet-Sauvignon adds structure and longevity to the blend.

The Saint-Ser domaine also provides a good example of microclimate. The altitude might appear a bit extreme—400 meters (1,330 feet)—but its coolness is balanced out by two compensating factors. One is the bright limestone cliff of Sainte-Victoire that acts like a solar furnace and focuses sunlight on the vines below. The other is the rubble shed off the steep slopes; it litters the vineyard and stores the heat provided by the sun during the day, radiating it back to the vines after dusk and prolonging the ripening process into the night.

Looking at Cézanne's paintings, we can see how the mountain lights up the landscape. What is more surprising is that vines rarely appear in his compositions. Probably there were fewer vineyards on the mountain in the late 1800s than today. Also, it was the mountain itself that fascinated the artist, not the details in the foreground, although his compositions are nonetheless exquisite. There is reason to believe, however, that Cézanne did not ignore vines altogether, and perhaps drew a bit of inspiration from them. As he wrote to his friend and author Émile Zola: "I used to love wine; now I love it even more. I have indulged in it; and I will indulge even more."

CHAPTER NINE

Languedoc's Vines
and Dinosaurs

The Languedoc-Roussillon region of southern France is a vast crescent stretching from the mouth of the Rhône valley westward to the Spanish border, framed by the Mediterranean Sea to the south and by an arc of mountainous segments to the north.

Such a natural, sun-drenched amphitheater was bound to attract farmers and vine growers, probably as early as the prehistoric Bronze Age. Historically, the onset of large-scale viticulture in Languedoc is attributed to the ancient Greeks, who planted vines at the mouth of the Hérault River (now the site of the city of Agde) during the fifth century BC.

The Romans took over control of the vineyards three centuries later. In 118 BC they created the Narbonensis province, which ran from the Alps to Languedoc, where its capital was established: the city and harbor of Narbo (Narbonne today). The wine production soared to such high levels that the province soon became a threat to the economy of Rome, leading Emperor Domitian (AD 51–96) to order the uprooting of Languedoc vines in AD 92.

When Emperor Probus (AD 232–282) reauthorized the culture of vines in AD 276, Languedoc reclaimed a leading position in winemaking, which it kept strengthening throughout the centuries. Chasselas, Muscat, and Carignan (the latter brought in from Spain) were among the first grape varieties that met with success. New ones were imported from the Middle East during the Crusades and experimented with on-site by the spreading monastic orders, which further broadened the diversity of Languedoc varietals and blends.

Well before the Crusades (1095–1291), monks had taken control of the vineyards. Their settlement throughout Languedoc was mostly the doing of Saint Benoît d'Aniane (750–821), who developed and reformed the Benedictine order in France and across Europe under the reigns of Charlemagne and of his son Louis the Pious.

Benoît d'Aniane has a remarkable life history. An aristocrat from birth, he rose to prominence in the Frankish court, becoming Charlemagne's official cupbearer and wine taster. At the age of twenty-three, he began a promising military career as an officer in the successful raid against the Lombards. The following year, traumatized by the near drowning of his brother (whom he saved by almost drowning himself), he decided to enter the religious life, and took his vows at the Saint-Seine Abbey in Burgundy. Appointed cellarer, in charge of cellar and kitchen supplies, his austerity did not go over well with his fellow monks. According to his biographer Arno, "Since with him wine did not flow freely, many monks looked at him sideways."

Considering the morals at Saint-Seine too loose, Benoît traveled to his native Languedoc and founded a monastery on his own land, on the banks of the Aniane River (a confluent of the Hérault). Soon he presided over hundreds of devoted monks, dispatching them across France to Aquitaine, Auvergne, and Burgundy to disseminate and enforce a stricter interpretation of Benedictine rules.

Along with religion, agricultural and winemaking know-how circulated as well, notably in Burgundy, where the Cluny order rose to power and ruled over the vineyards. In Languedoc, the monastery of Guellone, later renamed Saint-Guilhem-le-Désert, also rose to prominence, "surrounded on all sides by cloud-capped mountains," according to biographer Ardo, "and so full of assets that you would not wish a better place to serve God, for there are vines planted there, and the valley has many gardens."

Vines never ceased to prosper in Languedoc, especially since wine became a crucial trade currency in a country opening up to commerce. One monastery in particular, in the Aude valley at the foot of the Pyrenees, became highly successful at mastering a new type of wine that we will encounter when we make our rounds of Languedoc terroirs. Named Blanquette de Limoux, it is one of the first sparkling wines ever produced,

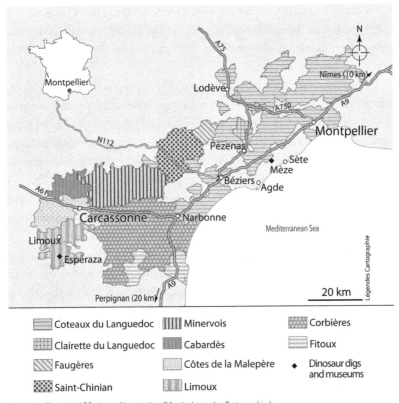

Figure 9.1 The main AOC wines of Languedoc. (Map by Legendes Cartographie.)

more than a century before Champagne. As for Languedoc as a whole, it has simply become the largest wine-producing region in the world, with 300,000 hectares (750,000 acres) of vines under culture today (fig. 9.1). A third of the wine produced in France, and 10 percent of the wine produced in the world, comes from Languedoc.

LANGUEDOC AND ITS MANY ASSETS

Shaped like a vast amphitheater facing the Mediterranean to the south and protected from the northern winds by the Massif Central, Languedoc is graced with an excellent climate. Sunshine is optimal and vines are kept dry, even along the coast, thanks to the winds that chase away morning fog—be they the *mistral* blowing out of the Rhône valley to the east, the *tramontane* funneling in from the west between the Massif Central and the Pyrenees, or the southwestern *cers* that the Romans worshiped to the point of assigning a god to it (Circius) because it rid the air of humidity and diseases such as malaria.

Under such favorable climes, it was tempting to plant vine everywhere when the wine market was on the upswing. Vineyards spread down the slopes and invaded the fertile coastal plains, yielding too many grapes and concentrating less aromatic molecules as a result. Languedoc winemakers today are moving back toward quality wine, keeping their production focused on the best terroir—the sun-drenched, well-drained hillslopes—and a small number of coastal vineyards that benefit from exceptional microclimates, such as Frontignan and Picpoul de Pinet on the outskirts of Montpellier, La Clape and Fitou near the city of Narbonne, and Collioure and Banyuls on the Spanish border.

Be it along the coast or inland, the terroir of Languedoc comes in many shapes and forms, with bedrock ranging from schist to sandstone, limestone, and clay, and different soil types often side by side in the same township.

The mountain ridges that buttress Languedoc to the north—the forest-covered Cévennes and Montagne Noire—are made of schist: old sediments, cooked and compressed into slaty layers. They go back 500 million years to the days when France was born out of the collision of tectonic plates, and are similar in age and origin to the schists of Anjou, which we saw in chapter 1.

Above the schist comes limestone deposited during the Jurassic and Cretaceous periods, when Languedoc was a shallow waterway connecting the nascent Atlantic Ocean to the west and the Tethys Sea to the east. The limestone is now raised high above sea level, forming the Causses Plateau on the edge of the Massif Central and the Corbières hills at the foot of the Pyrenees.

The Languedoc seaway shrunk during the Late Cretaceous as sea level dropped, and limestone sedimentation came to an end. Lakes and swamps took over, laying clay over limestone, and rivers dropped sand and gravel, recording the evolution of the landscape and building the varied terroir that would much later host the vineyards. We can trace these Late Cretaceous outcrops across Languedoc to the foot of the Pyrenees, discovering vineyards and dinosaurs along the way.

TRACKING DINOSAURS

Imagine Languedoc in the Late Cretaceous: floodplains crossed by meandering river channels, and swamps and lakes sparkling in the tropical sun, with tortoises and crocodiles snoozing on the mud banks and long-necked titanosaurs treading along the sandbars. Today, pockets of red clay and sand from the period outcrop near Mèze, 20 kilometers (12 miles) west of Montpellier.

Clay outcrops were first quarried in the Mèze basin over a century ago, to extract precious bauxite (aluminum ore). Dinosaur bones were unearthed in the process, catching the attention of paleontologists, who reclaimed the quarries once they were abandoned. Digs were organized on the most promising sites, competing for space with encroaching vineyards.

In order to view this remarkable cohabitation of vines and dinosaur digs, we drive past Mèze and its coastal oyster farms, and head up north on D 613 toward Montagnac, through a hummocky landscape of red clay crowned by pine trees and interspersed with vine lots. After a 5-kilometer (3-mile) drive, a turnoff to the right, marked by a small sign, leads to the Musée-Parc des Dinosaures. If visitors miss the turnoff, they will be warned at the next curve by a threatening, long-jawed *Spinosaurus*, staring at them from the top of a hill.

Open to the public, the fossil digs cover 5 hectares (12 acres) of pineland on the hillside. A nature trail takes you around the site, with display cases, information panels, and mounted dinosaur skeletons along the way. Around a corner, you will run into a couple of raptors hiding in the bushes (fig. 9.2); around another, you will admire a duck-billed *Hypacrosaurus* foraging along the path.

Several dinosaur species were found in the actual digs. Just north of the site, the fossilized bones of a new species of *Ankylosaurus*, christened *Struthiosaurus*, were unearthed. Right on-site, vertebrae of the parrot-beaked *Rhabdodon* were collected, as well as scores of teeth belonging to small carnivores, and whole nests of dinosaur eggs.

The fine sandstone and colorful clays also deliver excellent wines that gather their nutrients in this End-Cretaceous cemetery. Vineyards encircle the Mèze Dinosaur Park: we get back on route D 613, drive past the *Spinosaurus* on the hill, and after a few hundred meters (less than half a

Figure 9.2 Threatening, raptor-like dinosaurs haunted the forests of Languedoc during the Late Cretaceous, and shed their fossilized bones in many a terroir. (Photographed at the Mèze Dinosaur Park-Museum.)

mile) turn left into the side road leading to the Domaine Savary de Beau-regard.

Vines surround us as we drive up to the courtyard of the domaine's stone cottage. They cover 40 hectares (100 acres) of an ochre-colored soil that needs to be plowed in order for the clay not to squeeze and choke the vine roots. Winemaker Christophe Savary de Beauregard takes advantage of this wind-dried mineral terroir to grow a number of different grapes: five for the reds and rosés, and three for the whites.

Blending Cinsault, Syrah, and Grenache, the lightest red is a thirst-quenching *vin de soif* ("thirst-quenching wine") with a purple color and a bouquet of raspberry. Another blend mixes Merlot with Syrah and Gre-nache, resulting in a *vin de terroir* with a bouquet of black currant and prune, more generous on the palate. There is also a Cabernet-Sauvignon varietal, aged in oak barrels to last; over time, it will develop a bouquet of red berries and spices along with toasty notes.

The rosé wines of the estate team Grenache with Cinsault, delivering aromas of strawberry and red currant, and in another blend with a touch of Syrah, yielding aromas of peach and tangerine. As for the whites, they are produced as straight varietals: Christophe Savary offers a Chardonnay and a Sauvignon, but foremost Piquepoul, a local grape that has earned its own AOC appellation.

This grape has a history. It has been grown here for ages along the coastal Étang de Thau, a saltwater lagoon separated from the Mediterranean Sea by a 20 kilometer-long (12-mile) sandbar. Like other white grapes, Piquepoul is very receptive to the afternoon sea breeze that takes the edge off the afternoon sun, regulates chlorophyll production, and allows more time for the grape to reach maturity. Piquepoul yields a dry white wine known as Picpoul-de-Pinet (Picpoul is the spelling for the wine, Piquepoul for the grape). Light straw in color, it gives off a bouquet of peach, pear, and white flowers such as acacia, hawthorn, and lime tree. Fresh and supple, it is the perfect match for the oysters, sardines, and crustaceans of Mediterranean lagoons and other seas.

SAINT-CHINIAN: WINES AND DINOS

If we leave the fossil-rich basin of Mèze and keep driving inland to the north, up the Hérault valley, we leave the Cretaceous strata to enter older rocks; but this side trip is well worth it, since it brings us to visit remarkable monasteries and vineyards.

Saint-Saturnin, for instance, near Saint-Guilhem-le-Désert, produces a short-maceration wine called *vin d'une nuit* ("one-night wine"), since grape skins are left to macerate only overnight with the pressed juice. Light and fruity, Saint-Saturnin evokes cherry and raspberry with notes of cocoa.

Farther west, the Faugères vineyard covers hills of schist and produces sophisticated wines (see box). As for the Saint-Chinian vineyard, it straddles two geological provinces separated by a fault zone: old schists to the north, nearly 500 million years old, and the Late Cretaceous sandstone and limestone to the south, aged only 70 million years. Protected from the north wind by the Espinouse Massif, Saint-Chinian enjoys a warm and mild microclimate that favors grapevines as well as orange groves.

The Saint-Chinian vineyard produces two types of red wine that reflect the two different soils. To the north, the old, sandy schists break down into a lean, alkaline soil that evacuates rainwater and keeps the vine thirsty. Wines from this terroir, made from Mourvèdre, Grenache, and Syrah, are moderately strong, purple red, with aromas of red berries and

FAUGÈRES

On the upper tiers of the Languedoc amphitheater, the terroir consists of old marine sediments that date, in the area of Faugères, to the Devonian and Carboniferous periods 400 million to 300 million years ago.

The Faugères vineyard, 30 kilometers (19 miles) north of the coastal city of Béziers, covers the steep slopes of the Cévennes piedmont up to an altitude of 300 meters (1,000 feet), its terraces protected from erosion and slumping by stone walls. The Faugères schist breaks down into slabs of many colors, ranging from yellow ochre to orange, blue, and gray, and ultimately yielding a sandy soil that drains the water away. One of the great advantages of schist is that it soaks up solar heat during the daytime and radiates it back to the vines after sundown: winemakers are fond of saying that their grapes ripen overnight. The vineyard is protected from the north wind by the Espinouse and Caroux Massifs, and enjoys a mild winter and a hot and dry summer.

Faugères received AOC recognition in 1982 for red and rosé wines, and in 2005 for its whites. The winemakers are proud of their terroir, which gives the wine a distinctively mineral character, and their marketing strategy emphasizes the qualities and aesthetics of the colorful rocks. Blending any of the following grapes—Carignan, Cinsault, Grenache, Mourvèdre, and Syrah—Faugères wines are mostly reds (80 percent of the production) and are known to be full bodied, with a spicy bouquet and a good aging potential.

cherry pits, leather, and toasty notes. Once we pass the fault zone, south of the town of Saint-Chinian, the calcareous clay soils of the Late Cretaceous take over and yield wines that are stronger, with a bouquet of scrubland and aromatic herbs, and a long finish.

The southern half of the vineyard is a haven for fossil hunters, with dinosaur eggs and bones buried in its clay and sandstone. One of the best dig sites is Cruzy, 10 kilometers (6 miles) south of Saint-Chinian, where an important, albeit serendipitous, discovery was made in the 1980s.

Collecting eggshell fragments in the vineyard is relatively easy, if you know where to look and what to look for. On the other hand, finding huge bones from an adult dinosaur is a rare event. Winemaker Jean-Claude Guilhaumon ran across two convenient round-ended rocks that he used as props to block the wheels of his tractor—until he realized that they resembled giant femur heads and signaled his find to paleontologists Eric Buffetaut and Jean Le Lœuff. Dinosaur bones they were, and the winemaker turned over several acres of his vineyard to the scientists, who organized a fossil dig.

Over the years, the Cruzy site has delivered a full End-Cretaceous menagerie, including fish; frogs and salamanders; tortoises; several types of crocodile (some with sharp teeth, others with crunching teeth); a Komodo dragon–like reptile; a flying reptile and a primitive bird; and a variety of dinosaurs.

With respect to the latter, the Saint-Chinian vineyard has yielded bone fragments of giant vegetarian quadrupeds of the Titanosaurus family; of the smaller, parrot-beaked *Rhabdodon*; and of carnivorous bipedal dinosaurs, including the giant Abelisaurid individual that left its hind legs as a souvenir on the estate of Jean-Claude Guilhaumon.

The reigning star of Languedoc dinosaurs, however, is to be found farther west in the Aude valley, south of the city of Carcassonne. Not only is the valley a treasure trove for fossil hunters, but it also hosts a little-known secret for wine connoisseurs: its Benedictine monastery invented and developed a sparkling wine nearly two centuries before Champagne rose to fame.

THE SECRET OF SAINT-HILAIRE

To seek out dinosaurs and sparkling wine, we leave the Mediterranean coast and head west in the direction of Toulouse, following the corridor framed by the Montagne Noire to the north and the Corbières hills to the south. After a 50-kilometer (30-mile) drive, we reach the walled city of Carcassonne, overlooking the Aude River and its vineyards. At this point we turn south on route D 118, driving up the Aude valley toward the Pyrenees.

The mountain range did not start to form until the end of the Cretaceous, when the Iberian microcontinent swung counterclockwise to pinch the stack of sediments lying between Spain and France. As the elevation began to build, torrents rushed down the slopes and filled the Languedoc trough below it with cobbles, gravel, and sand. The bedrock underlying the medieval city of Carcassonne, which provided the building blocks for its ramparts and monuments, is a natural form of concrete, made of rubble from the Pyrenees and cemented together by greenish clay.

If we wish to reach the Cretaceous strata that correspond to the last days of the dinosaurs, we need to climb out of Carcassone's elegant green molasse—the clay-rich, sedimentary rock—and head up the Aude valley that digs through the older rocks in its upstream reaches. Here, heading up the river means heading back in time.

Fifteen kilometers (9 miles) south of Carcassonne, we are still driving through young Cenozoic rocks, having not yet reached the Cretaceous, but we momentarily interrupt our time travel to take a side road east of the river and head up to the Abbey of Saint-Hilaire: the cradle of sparkling wine.

Saint-Hilaire (sixth century AD) was the first bishop of Carcassonne. Legend has it that he built his first chapel in the remote location that now bears his name. The abbey was later built by Benedictine monks on the sacred site under the reign of Charlemagne, and is mentioned for the first time in AD 825. Its long history is marked by wars, pillaging, destruction, and bold architectural restoration that introduced new styles using both golden limestone and green molasse from Carcassonne. The walls of the abbey enclose the monastic buildings as well as the village houses under its protection.

The Saint-Hilaire cloister is now a public garden, framed by slender double columns. The refectory is famous for its acoustic design, with a vaulted preaching balcony or "reading chair" high above the floor, from which the orator's voice floated down mystically upon the congregated monks. Today, the refectory hosts lectures and concerts. But of most interest to us is the abbey's cellar, where an important discovery was made around 1531, according to monastic records.

The region had been famous for its white wine. Already in the Middle Ages, court historian Jean Froissart (1337–1404) described it as being particularly "delectable." But by the early sixteenth century, the monks of Saint-Hilaire were running into a difficult problem. They had begun storing the wine in glass bottles, rather than in barrels and jugs, and closing them carefully with wooden corks and beeswax seals. To their dismay, after several months of storage, the bottles would violently eject their cork or even shatter to pieces. Accidently, the monks had run across the phenomenon of secondary fermentation and carbon dioxide accumulation inside a closed bottle—the recipe of sparkling wine, or Blanquette de Limoux, as it became known in the region.

More than a century would pass before another Benedictine monk by the name of Dom Pérignon (1639–1715) ran into the same problem in his Champagne Abbey of Hautvillers; he, too, mastered it, producing sparkling wine. Yet the Aude valley and its Abbey of Saint-Hilaire proudly claim to be the inventors of the sparkling method.*

BLANQUETTE SPARKLING WINE

There are different ways to make sparkling wine in the Aude valley. In the so-called ancestral method (*méthode ancestrale*), the Blanquette de Limoux is made with Mauzac white grape. This variety comes from Gaillac, northeast of Toulouse, where it is also used nowadays to make sparkling wine, as well as a sweet white. The grape is naturally resistant to mildew, well adapted to limestone and clay soil, and a late bloomer that matures slowly and takes time to concentrate aromas—typically a bouquet of ripe apple, quince, and candied fruit.

In the ancestral method, the natural sugar and yeast contained in the grapes are deemed sufficient to ensure fermentation, which lasts from harvest until the month of March, when the wine reaches 6 or 7 degrees proof. It is then bottled, and undergoes a second *in vitro* fermentation during the spring: the transformation of malic acid into lactic acid, accompanied by the release of carbon dioxide that generates the famous bubbles. The sparkling wine obtained in this manner is light, low in alcohol, and slightly sweet.

The alternative traditional method (*méthode traditionnelle*) involves further steps in the winemaking process. First of all, Mauzac is not the exclusive grape, although it is overwhelmingly dominant (over 90 percent). Chenin and Chardonnay can account for up to 10 percent of the blend. The varieties are vinified separately, then blended according to the whim and aromatic objectives of the winemaker before being bottled. At this stage, as is also done in Champagne, a small amount of sugary, yeast-rich "liquor" is added to the wine to foster the second round of fermentation and increase the production of carbon dioxide. The bottles are stored on racks, bottoms up, for nine months; then they are uncorked to expel any undesirable yeast deposits. To make up for the lost volume, the bottles are topped off with a new dose of "expedition liquor," which defines the sweetness of the finished product according to its sugar content: brut or

---

*Other regions in France also make the same claim, namely the region of Die, north of Avignon in the Rhône valley, that allegedly developed its sparkling wine in Roman times (Clairette de Die).

semidry (*demi-sec*). Finally, the bottles are laid down to rest nine more months before hitting the market.

As for the terroir of the Blanquette de Limoux—over 2,000 hectares (5,000 acres) authorized under its AOC status—it straddles four different climate zones. In order to highlight this interesting diversity, Limoux winemakers decided to plant rosebushes of different colors alongside their vines, according to each zone. Red roses, for instance, are planted in the Mediterranean zone, which is characterized by a warm climate and a regular distribution of humidity and rainfall provided by a dominant sea breeze. Grapes mature on the early side and are the first to be harvested.

Going up the Aude valley, the second zone is named the Autan terroir, on the hills encircling the city of Limoux; it displays yellow roses. The elevation is still less than 200 meters (650 feet), and the vines are protected from the sea breeze by the Corbières hills, resulting in a warm and dry climate. Autan grapes are harvested second.

West of Limoux is the third terroir, signaled by pink roses: the hills of Chalabrais and Malepère on higher ground (200 to 300 meters [650 to 1,000 feet]), which begin to feel the influence of oceanic air spilling in from the Atlantic. The vines are interspersed with groves of beech and oak trees, and their ripening and harvesting naturally come later.

But the most interesting terroir, at least for geologists, is known as the High Valley (Haute Vallée), up the Aude River toward the Pyrenees and above 300 meters in elevation. Bearing orange roses, the terroir is set on End Cretaceous limestone (fig. 9.3) and sandstone that were deposited when the area was covered with lakes, floodplains, and river channels, and home to some of the last species of dinosaurs to roam the Earth, a few million years before their brutal extinction.

The village of Espéraza is the gateway to this sanctuary. Past the stone bridge over the river, up near the old train station, a dinosaur museum and research laboratory are open to the public.[†] Two kilometers (1¼ miles)

[†] Musée des Dinosaures, 1 place du Maréchal de Lattre de Tassigny, 11260 Espéraza. Open from 10:00 a.m. to 7:00 p.m. in July and August (10:30 a.m.–12:30 p.m. and 1:30–5:30 p.m. the rest of the year). In the summertime, the price of admission ticket includes access to the Bellevue dinosaur digs down the road at Campagne-sur-Aude. Phone: 04 68 74 26 88; website: www.dinosauria.org/.

Figure 9.3 The vineyard of Blanquette de Limoux AOC sparkling wine covers the slope uphill of the village of Campagne-sur-Aude. The cliff in the background displays the last limestone and clay strata of the Cretaceous period.

further upstream, we reach the even smaller village of Campagne-sur-Aude, where vineyards and dinosaur bones overlap in a very special way.

CUVÉE OF THE DINOSAUR

After crossing the Aude River, the street around the church at the back of the village leads to a stately white building with a terrace and a fancy balustrade—very Spanish looking—and a courtyard with loading docks and cases of wine ready to go. These are the headquarters of the Salasar family, established at Campagne-sur-Aude since 1890, ruling over 200 hectares (500 acres) of vines, and producing white, red, rosé, and sparkling wines, including of course the famous Blanquette de Limoux.

If visiting in the late afternoon, just before closing when work in the vineyard has ended, you might run into René Salasar himself, third generation of the founding family, who will pop open a few bottles and run you through the whole gamut of his production. He will have you start with his still wines—the unsparkling ones—such his Pays d'Oc Pinot Noir, his Chardonnay that gives off a bouquet of peach and almond, and his rosé leaning toward cherry and bell pepper.

Sparkling-wise, René Salasar offers a Blanquette de Limoux brut for cocktail hour or to accompany shellfish; a honey-colored Blanquette semidry to go with dessert; or a sweet Blanquette for teatime and pastry. Besides Blanquette, the winemaker produces another sparkling wine named Crémant de Limoux that also does very well in wine-tasting competitions, and in which the Mauzac grape is replaced by Chardonnay and Chenin Blanc.

All these grape varieties are grown in small lots perched on the steep slopes of the Aude valley. The spread in their locations is good insurance against localized hailstorms that might wipe out one vine lot but not all: the winemaker's version of the proverbial "do not put all your eggs in one basket." But one lot in particular gets special attention, as it yields the starring wine of the estate: La Cuvée des Dinosaures ("The Dinosaur Cuvée").

The lot in question is Bellevue, up the hill just behind the Salasar headquarters. On a clay-rich, sandy slope high above the village, vine growers had long noticed that the rocks there seemed to contain "something." Paleontologists were quick to identify these "somethings" as fossils. They extracted the first dinosaur bones at Bellevue in 1993, under the supervision of Eric Buffetaut and Jean Le Lœuff, who established their headquarters at Espéraza and founded the Dinosaur Museum there.

DINOSAUR OF THE VINES

On the site where the Blanquette's Mauzac grape once grew, the dig teams uncovered one of the best-preserved dinosaur specimens in Europe: a nearly complete skeleton of a 12-meter-long (40-foot) titanosaur. On the basis of the recovered bones, it was found to be a new species living in France in the End Cretaceous. In recognition of the unusual site where it was found, it was named *Ampelosaurus atacis*, which is Latin for "Dinosaur of the vines from the Aude area" (atacis is the Aude area tribal name). Its long-necked, long-tailed silhouette now decorates the label of René Salasar's special Dinosaur cuvées (fig. 9.4).

The terroir of Bellevue and its fossil beds tell a remarkably detailed story. Fine sandstone and claystone point to a meandering riverbed at the time, in the midst of an alluvial plain. Palm trees and palm-like cycads clustered along the banks, fish resembling pike swam the river channels

Figure 9.4 The Salasar estate at Campagne-sur-Aude named one of its top cuvées of Blanquette de Limoux for the *Ampelosaurus atacis* dinosaur discovered on one of its vine lots.

(their fossilized scales were found), and reptiles came there to drink. The dig teams recovered fragments of tortoise shells, crocodile teeth, and of course the famous dinosaur bones themselves.

Many different dinosaur species lived and died on the floodplains. The search teams have found at Bellevue the remains of armored Ankylosaurs, of *Rhabdodon* ornithopods—those parrot-beaked herbivores that we already ran across in Provence—and of carnivorous theropods, including the small and ferocious *Variraptor* and the large *Tarascosaurus*—a Franco-African version of America's T. rex that shed a few dagger-shaped teeth on the site.

As for *Ampelosaurus atacis*, the giant herbivore with a long neck and a long tail, apparently several were entombed at Bellevue, including an almost complete 12-meter-long (40-foot) specimen that has risen to stardom. It is even known by its first name, Eva, in honor of student Eva Morvan, who discovered its first bone during the 2001 dig season.

After five summers of hard and patient work, the dig teams recovered most of the animal's bones, including a good deal of its brain case, which is particularly rewarding in view of how thin the cranium happens to be. Based on all the gathered remains, Eva was apparently an adolescent—

12 meters long, whereas an adult probably reached 18 meters (60 feet), according to other bones found on-site. She met an untimely death, perhaps drowning in a deep spot of the river, and her carcass came to rest in a river bend, where it was quickly covered up by sand and silt—a tomb that would lie unopened for 70 million years.

Ampelosaurus atacis is one of the most exciting paleontological discoveries ever made in the French vineyards. The Bellevue lot that was once entirely devoted to Mauzac grapes and Blanquette de Limoux—you can see old vine roots sticking out of the rock face right above the digs—has depicted for us a scene of life and death in the Cretaceous, 70 million years ago. And when, at the end of a hot summer day spent digging and clearing fossils, paleontologists and their student volunteers come down to René Salasar's cellar for a well-deserved Blanquette, the streams of bubbles that rise along the glasses might bring to mind the dying gasps of the last French dinosaurs.

THE END OF THE DINOSAURS

Provence's Montagne Sainte-Victoire, the hills of Saint-Chinian, and the upper Aude valley are not the youngest sites in France to have yielded dinosaur fossils. A few sites are ever closer in time to the 65.5-million-year dateline of the dinosaurs' brutal mass extinction: they are located in the Haute-Garonne district, close to the Pyrenees Mountains, and at Fontjoncouse, on the Mediterranean-facing flanks of the Corbières hills.

On this land broken up into a mosaic of hills and scrubland, cut through by numerous gullies, the stone walls of sheep pens point to a grazing tradition rather than to agriculture. Only vines and olive trees can grow on this arid limestone and sandstone terroir, interspersed with pockets or red clay. Carignan, Grenache, and Syrah grapes yield dense and strong red wines, and fresh, tangy rosés.

A short distance from the vineyards, Éric Buffetaut, Jean Le Lœuff, and their team from the Espéraza Dinosaur Museum have searched the terminal strata of the End Cretaceous. The fossils they uncovered do not belong to the great titanosaurs or the ornithopod *Rhabdodon*, as was the case in the slightly older strata of the Aude valley, but to duck-billed dinosaurs of the Hadrosaur family, and namely to the species *Telmatosaurus*. Since the grater-shaped teeth of the hadrosaurs were adapted to particularly tough vegetation, it appears that the regional climate turned harsher at the very end of the Cretaceous, and that the change in vegetation resulted in a different cast of dinosaur species.

The rise of hadrosaurs to a near-monopoly in southern France at the close of the Cretaceous does not mean that the worldwide diversity of dinosaurs was declining at the time. In southern France, there are fossilized remains of other species besides the hadrosaurs, namely of carnivorous dromaeosaurs ("raptors") at Fontjoncouse, and of an ankylosaur species in Haute-Garonne, as well as the tracks of a titanosaur on the other side of the border, in Spain.

In fact, about a dozen different species of dinosaurs are identified throughout Europe in the very last strata of the Cretaceous: the Maastrichtian stage (dated 70.5 to 65.5 million years ago). In the American West, where the time period is particularly well studied, no less than 40 dinosaur species are tallied. Basing his estimate on the provincial distribution of species in ecosystems, Jean Le Lœuff extrapolates that anywhere between 600 and 1,100 different dinosaur species roamed the Earth at the very end of the Cretaceous.

The picture that emerges is that the diversity of animal species did not decline progressively, but flourished in the Late Cretaceous, up to a moment in time when it was brutally interrupted by a great mass extinction 65.5 million years ago, when 75 percent of all terrestrial and marine species disappeared from the face of the Earth. This dramatic collapse of the ecosystem is today convincingly tied to the collision of a 10-kilometer-sized (6-mile) asteroid with our planet. Numerous proofs of the collision are found in the last inch of clay of the Cretaceous, including minerals shocked by cosmic-speed impact pressures and molten glass sprayed by the blast. The resulting crater, 200 kilometers (125 miles) in diameter, was identified below the surface of Mexico's Yucatan peninsula. Other theories put forth to explain the mass extinction (climate change due to volcanism; marine ecological stress due to a sea-level drop) lack evidence and receive little support today.[‡]

[‡] For a full perspective on the end of the dinosaurs and the research that led to the discovery of an asteroid collision with Earth, see Charles Frankel, *The End of the Dinosaurs: Chicxulub Crater and Mass Extinctions* (Cambridge University Press, 1999).

CHAPTER TEN

Champagne

"The dinosaurs are dead! Long live the mammals! Let's have Champagne!" some of us may be tempted to say—and it so happens that the Champagne region bridges the end of the Cretaceous and the dawn of this new era, registering in its Early Cenozoic strata the rise and diversification of mammals that ultimately led to sophisticated primates and *in fine* to the human species. Paleontologists comb through the sands and clays at the foot of the Reims Mountain to collect the tiny teeth and bones of our distant ratlike ancestors.

Of course, it isn't mammal teeth that have made the Champagne region famous, but its vineyard (fig. 10.1, top) that straddles the boundary between the Cretaceous and the Cenozoic and produces the most famous, festive, and expensive sparkling wine in the world.

Such a success was in no way preordained. This region belongs to the Paris Basin, and by definition a basin tends to be flat and ill suited to vine growing. In addition, its northerly location and cool climate were major drawbacks, even though the proximity to Paris and its wine-buying market was a clear advantage.

What transformed a flat and cool terrain into an illustrious wine terroir has much to do with regional structural geology. Over the last few million years, the uplift of the Alps and by domino effect of the Vosges Massif to the north raised the eastern edge of the Paris Basin. Because they are now tilted, the stacked sedimentary strata are beveled by erosion into a series of steps, each step consisting of a cliff or cuesta facing the rising sun (fig. 10.1, bottom). Otherwise a monotonous plain suited only to

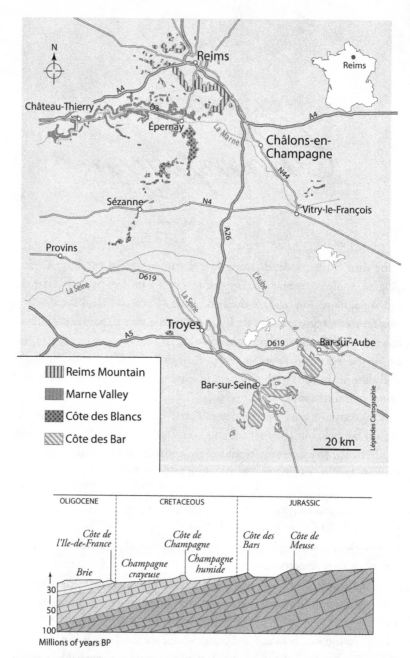

Figure 10.1 *Top*, the Champagne AOC vineyard occupies the slopes of several escarpments or cuestas of the Paris Basin. *Bottom*, the cross section through the basin shows the pile of tilted sedimentary strata—the hardest strata form the escarpments. *BP* signifies before present. (Map by Legendes Cartographie.)

extensive agriculture (wheat and sugar beet), the Champagne region is thus graced with a few sloping areas that receive solar rays at nearly a right angle and make up somewhat for the northerly location.

For this reason, vines were planted early on in the region, probably by the third or fourth century AD. Pinot Noir and Chardonnay became the leading grapes and delivered red and white wines that competed with those of Burgundy and the Loire valley, taking advantage of Champagne's proximity to Paris and of the navigable Marne River—confluent of the Seine—for transport and delivery.

Credit for this success should also be given to the monasteries and powerful dioceses of Reims and Châlons-sur-Marne, and to the fact that the crowning of French kings traditionally took place in the Reims cathedral: the wines served during these regal events were naturally Champagne wines, in particular the excellent Bouzy reds.

The Sun King, Louis XIV (1638–1715), was notoriously fond of Champagne wines, to the point where Burgundy took offense. A fierce rivalry broke out between the two provinces, each one vying for the favors of the king and his court, and claiming superiority over the other. The medical bodies of Paris and Reims took sides in the dispute, since the health of the monarch was evidently at stake. Despite the warnings of his official physician, Guy-Crescent Fagon, who sided with the Burgundy wine merchants, the Sun King never repudiated Champagne wines—or the wines of other French provinces, for that matter.

It is also under the reign of Louis XIV that the wines of Champagne took an astonishing and sparkling turn for the better, under the guidance of Benedictine monk Dom Pérignon (1639–1715), cellarer of the Abbey of Saint-Pierre d'Hautvillers.

Contrary to the legend, it is not Dom Pérignon who invented sparkling wine. As we saw in the last chapter, credit for the discovery can be given to the Abbey of Saint-Hilaire in Languedoc, more than a century earlier, or even to other areas of France, such as Die in the Southern Alps. Although there is no proof that Dom Pérignon actually traveled to Languedoc and learned the method from his Benedictine brothers, some technical know-how might have spilled over, so to speak, from one monastery to the other. According to other sources, the process was fine-tuned in England, albeit by meddling with imported Champagne wine, and pressure was put on Dom Pérignon to perfect the method for the English market.

Be that as it may, once Dom Pérignon set his mind to mastering the sparkling process, he made it into an art form that became known as the *méthode champenoise*. The basic process, which also applies to the Blanquette de Limoux and other sparkling wines, consists in having the wine undergo a second round of fermentation once it is trapped inside a corked bottle: residual sugars are metabolized by the remaining bacteria, releasing carbon dioxide in the liquid and in the air space beneath the cork.

Mastering the process is no easy matter, since the pressure inside the bottle rises to 5 or 6 bars—as much as a scuba diver experiences at a depth of 50 meters (over 160 feet) of water—and in those experimental days, one out of three bottles would end up blowing its cork or, worse yet, shattering to pieces. In order to control this "devil's wine," as it came to be known, Dom Pérignon modified the shape of the bottle, increased the glass thickness, and improved the cork's fastening by way of a wire cage.

Dom Pérignon also developed the philosophy of assembly, vinifying grape varieties separately (as well as batches of the same grape coming from different terroirs) before carefully blending them in the right proportions to combine qualities, erase defects, and achieve the best result possible. Finally, when it came to storage, the cellarer dug deeper cellars under his abbey, conscious that cool, constant temperature and humidity helped the aging process.

Legend has it that sparkling Champagne first came to the attention of the court of Louis XIV when it was purportedly introduced by a charming baroness by the name of Jeanne de Thierzy. What is established is that it was very popular at the court of the next king, Louis XV, and in particular was a favorite of his mistress, Madame de Pompadour. Champagne wineshops opened up in Reims and Épernay, and its reputation spread across Europe and Russia, thanks to the connections of aristocratic vineyard owners who traveled widely to promote their sparkling wine.

Champagne is entrusted with many virtues. Besides aiding digestion, it is believed to have anti-inflammatory properties, owing to its natural richness in potassium, calcium, and magnesium. Even before the advent of modern medicine and psychiatry, Champagne was known to fight anxiety and depression, a property attributed today to its trace amounts of zinc and lithium.

In other words, Champagne is irreplaceable. As estate owner Elizabeth Bollinger (1899–1977) wittily declared to the London *Daily Mail* in 1961:

"I drink it when I am happy and when I'm sad. Sometimes I drink it when I'm alone. When I have company I consider it obligatory. I trifle with it if I'm not hungry and drink it when I am. Otherwise I never touch it—unless I'm thirsty."

As for Emperor Napoleon Bonaparte (1769–1821), he was quoted as saying, "I cannot live without Champagne. In case of victory I deserve it; in case of defeat I need it."

CHAMPAGNE AND TERROIR

Aside from its distinctive production method, its bubbles, and its many virtues, Champagne would not be Champagne were it not for its unique terroir. Local winemakers have always been conscious of this, and have fought hard to defend their heritage and name. Faced with growing competition from other sparkling wines using roughly the same vinification techniques and calling their products Champagne, they sought and obtained, as early as 1887 from the Angers Court of Appeals, a judgment limiting the use of the word *Champagne* solely to wines from their region. This Controlled Appellation of Origin (AOC) applies to 30,000 hectares (75,000 acres) of vines today.

The Champagne vineyard is established on the concentric ridgelines—called *côtes* or cuestas—that run through the Paris Basin and mark the transitions between different sets of sedimentary strata, piled up in a bull's-eye pattern. As we drive out of Paris and head east across the sedimentary basin, the first cliff-like transition we reach separates the upper, younger layers of the Cenozoic (the age of the mammals) from the older Cretaceous sediments to the east that constitute the Champagne agricultural plains. This steeply sloping transition is known as the cuesta or Côte de l'Île-de-France.

Vines are planted along the slope of an escarpment that faces mostly east, but is also scalloped so that it faces other directions as well. It boasts in particular the Montagne de Reims ("Reims Mountain"), a major promontory overlooking the city of Reims in the Cretaceous plains below. It is bounded to the south by the Marne River, which cuts the cuesta at a right angle and carves its way westward through the plateau. South of the Marne valley, which offers an extra set of vine-growing slopes, the cuesta resumes its southward direction, lining up vineyards over close to 100 kilometers (60 miles) in two long segments known as La Côte des Blancs and La Côte de Sézanne.

If we keep traveling eastward across Champagne, we leave the vine-covered cuesta of Île-de-France for the flat Upper Cretaceous limestone plains—the land of wheat and sugar beet—and need to drive 50 kilometers (30 miles) before reaching the next scarp dropping down one more step to the Lower Cretaceous strata—a cuesta known as La Côte de Champagne. The name refers to the broad region of Champagne and not to its wine, for there is not enough of a slope here to plant good vines except in the area of Vitry-le-François. Moreover, the Lower Cretaceous chalk has a high clay content, which makes the soil rather impermeable and soggy, earning the region the name of Champagne Humide ("Damp Champagne").

Still father east, a third scarp is La Côte des Bar, named for its principal townships: Bar-sur-Aube at its northern end and Bar-sur-Seine, 40 kilometers (25 miles) to the southwest. Here the slopes are sufficiently marked to plant vines again—a profile due to hard Jurassic strata that break almost vertically and build up a steep talus. La Côte des Bar constitutes a separate, southwesterly terroir that also is authorized to market its sparkling wine as Champagne (see box).

IN PRAISE OF CHALK

The Côte de l'Île-de-France, overlooking Reims and Épernay, offers the best conditions for vine growing, be they slope or climate, and also the most typical terroir, derived mostly from chalk.

Chalk is a special form of limestone, soft and porous, that was built from carbonate material that accumulated on the seafloor in the Mid Cretaceous (around 100 million years ago). Its carbonate grains are the outer skeletons, or tests, of large plankton that blossomed in the tropical seas covering France at the time.

Now exhumed, this white, permeable limestone breaks down easily and so is of little interest to architects and masons. The beautiful Gothic cathedral of Reims, for instance, is built of younger, harder limestone. Nonetheless, the more modest churches of Châlons-sur-Marne and small Champagne villages rely on this local chalk, which is still mined at La Veuve on the outskirts of Châlons.

CÔTE DES BAR AND ROSÉ DES RICEYS

La Côte des Bar is located in the department of Aube, 150 kilometers (under 100 miles) southeast of Reims, and stretches between the towns of Bar-sur-Aube and Bar-sur-Seine over a distance of 40 kilometers (25 miles). Its ridge marks the transition between End Jurassic resistant limestone (Portlandian stage, 145 million to 150 million years ago) and older and softer marly limestone (Upper Kimmeridgian stage, 150 million to 155 million years ago) that form the lower slopes and plains below.

La Côte des Bar is well suited to vine growing, especially the steepest slopes facing south-west to southeast along the scalloped boundary carved by the tributaries of the Aube and Seine Rivers. Limestone rubble covering the slopes reflects sunlight and heats the vines and soil while efficiently draining excess rainfall and regulating the amount of water available to the vines.

The calcareous nature of the soil favors Pinot Noir over Chardonnay. Pinot Noir accounts for 87 percent of the Côte des Bar vineyard today, and probably was just as prevalent in the past, since the Côte des Bar has traditionally produced red and rosé wine. The latter is still produced today under the renowned appellation Rosé des Riceys: its reputation goes back to the late seventeenth century, when masons from the region were hired to work on the Château de Versailles and brought barrels of wine with them. Supervising the construction of his palace, the Sun King, Louis XIV, who decidedly was very curious when it came to wine, tasted the Côte des Bar wine and found it excellent. Rosé des Riceys has a bouquet of hawthorn, violet, raspberry, vanilla, and hazelnut.

Côte des Bar winemakers also launched into the production of sparkling wine, imitating their successful neighbors of Reims and Épernay, but these regions have long barred them from using the name Champagne on their bottles. It took an uprising by local producers in 1911 before they were granted rights to the name, and La Côte des Bar now accounts for a quarter of total Champagne production. It grows mostly Pinot Noir (87 percent), with approximately 7 percent of Chardonnay, 5 percent of Pinot Meunier, and a few hectares of Pinot Blanc thrown in for good balance.

In terms of agriculture also, chalk is a rather poor material low in nutrients, which forces grapevines to extend their root system and to tap a large area of rock, reaching downward through cracks or along horizontal planes. Chemical-wise, chalk provides some calcium and magnesium to the soil, but little more. On the other hand, the material is permeable, so rainwater sinks through, which keeps the root systems relatively dry. This property of the limestone also allows the roots to tap humidity at depth in the heat of the summer, through the capillary ascent of water droplets back toward the surface. Chalk acts therefore as a water regulator throughout the year.

Three grape varieties in particular profit from this low fertility, good water circulation, and relatively cool climate: Chardonnay, Pinot Noir, and Pinot Meunier. Champagne is a blend of these, in proportions that vary according to place, vintage, and character sought by the winemaker. Since grape juice is essentially colorless when it is kept from macerating with grape skins, one can make clear-colored wine with black grapes such as Pinot Noir and Pinot Meunier: the resulting Champagne is described as a "white from reds" (*blanc de rouges*). At the other end of the spectrum, Champagne can be made from white Chardonnay grapes alone, as is done in the aptly named Côte des Blancs ("Hillslope of Whites"; see box) south of Épernay, and it is then described as a "white from whites" (*blanc de blancs*). In between, most Champagne is a blend of all three varieties.

The prevalence of one grape variety over another depends on the terroir, both soil-wise and climate-wise: Chardonnay in the Côte des Blancs and its southern extension (Côte de Sézanne); Pinot Meunier in the Marne valley; and Pinot Noir in the Côte des Bar and on the Reims Mountain—this last also hosting Chardonnay and Pinot Meunier on account of its range of soil types, slopes, and exposures.

CÔTE DES BLANCS: CHARDONNAY CHAMPAGNE

South of the Marne valley and the city of Épernay, the Côte de l'Île-de-France takes on the name of Côte des Blancs ("Hill Slope of the Whites") because of the predominance of the Chardonnay grape variety, accounting for 95 percent of the vineyard. This white grape loves limestone and is rooted in a very pure form of chalk from the Late Cretaceous Campanian stage (84 million to 70 million years ago). In its higher reaches, the vineyard taps limestone rich in clay belonging to the younger Lutetian stage (49 million to 40 million years ago).

Despite the relatively uniform, chalky terroir, there are slight differences in taste from one Champagne to the next. Those from the township of Cravant to the north are stronger than those from Avize in the center, which are more delicate and longer on the palate. Those from Oger in the south are the most delicate.

THE REIMS MOUNTAIN

The Reims Mountain is a "cape" that sticks out from the north–south-trending hill front of the Côte de l'Île-de-France, bounded to the north by the Vesle River and to the south by the Marne valley (fig. 10.2, top). Half an hour west of the city of Reims, a small road known as La Route des vins

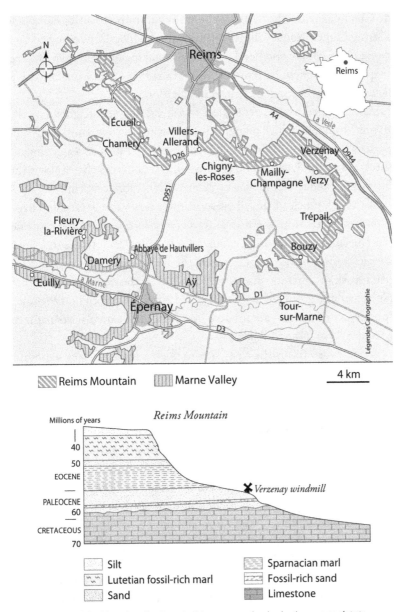

Figure 10.2 *Top*, the Reims Mountain and its vineyards. *Bottom*, cross section showing the sequence of strata—limestone, marl, and sand—ranging from the End Cretaceous (approximately 70 million years ago) to the Late Eocene (approximately 40 million years ago). (Map by Legendes Cartographie.)

de Champagne climbs into the vineyards and runs cross slope through a string of picturesque villages. Higher up, the hillcrest is crowned by a dense forest of oak and beech—the beech taking on a peculiar twisted shape in the highest, easternmost part of the forest (Les Faux de Verzy), believed to be the result of some odd mutation.

An estimated 9,000 hectares (22,500 acres) of vines are grown on the Reims Mountain, which amounts to approximately one-third of Champagne's total vineyard. In its upper reaches, the hill is also a geological showcase of the Early Cenozoic era—the time frame that follows the Cretaceous and its terminal mass extinction (fig. 10.2, bottom).

Although much of the vineyard is lower down in Cretaceous chalk, it profits greatly from these upper layers of Cenozoic marl and clay that crumble and wash downslope, mixing extra minerals into the soil. Such a varied terroir fits all grapes, so that the Reims Mountain vineyard is a patchwork of different shades of green leaves: the dark green of Pinot Noir; medium green of Chardonnay; and lighter green of Pinot Meunier, speckled with white—a flour-like sprinkling that earned it its name of Meunier ("From the Mill").

It comes as no surprise that with so many variables—soil, exposure, and grape variety—the Reims Mountain hosts as many different Champagnes as there are villages, and there are nearly a hundred villages around the hill. Some of the townships have earned Premier Cru or even Grand Cru status in the same way that Bordeaux and Burgundy villages have, in recognition of the excellence of the terroir and the resulting quality of the wines. The best way to grasp such range and variety is to follow the touristic itinerary *La Route des vins de Champagne* around the hill.

Leaving the city of Reims, we first head west across the plain until we reach the northwestern foot of the mountain. There we hook up with scenic route D 26, which goes through Coulommes-la-Montagne, Pargny, Jouy-lès-Reims, Ville-Dommange, Sacy, Écueil, and Chamery, to list but a few, along the scalloped hill front. Cenozoic layers of sand, marl, and clay contribute greatly to the terroir, and the northeastern exposure is a bit on the cool side, which suits Pinot Meunier better than its two partners.

There is a subtle change in terroir as we cross route D 951—the road that runs north to south from Reims to Épernay and divides the mountain into two halves. As we enter the eastern half, passing through the villages of Villers-Allerand, Rilly-la-Montagne, and Chigny-les-Roses, the higher slopes are still rich in clay, but lower down the vineyard spreads out on a gently sloping sandy terrace that covers the Cretaceous chalk. On this more calcareous terroir, Pinot Meunier progressively gives way to Pinot Noir.

One of the prominent landmarks of this northern half of the mountain is the Verzenay windmill (fig. 10.3), its wings now still, facing east across the Champagne plain—such an exceptional view that it was used during World War I as an observation post by the Allied forces and was slightly damaged by a German shell fired from across the plains.

The terroir's sandy limestone belongs to the Paleocene epoch, 65.5 million to 56 million years ago—that is to say, to the first stage of the Cenozoic period, right after the mass extinction of the dinosaurs. At the time, the Paris Basin was still flooded by a tropical sea, but the water level was oscillating and the shoreline moving back and forth across the site of the future Champagne vineyards. At one point the coastline passed through the Reims area, with an estuary fitting the location of today's village of Cernay. Paleocene sediments here contain fossils of crocodiles, tortoises, giant running birds, and a wide variety of mammals: hedgehog-

Figure 10.3 The Verzenay windmill towers over the vineyards of the Reims Mountain's northern slope.

like insectivores; hyena-like carnivores; early ungulates that looked like miniature horses; and even lemuroid species that opened up the way for more primates to come.

Sediments from this Paleocene period underlie most of the Verzenay vineyard, but vines in the highest reaches grow on younger sands and clays that belong to the second epoch of the Cenozoic mammal era called the Eocene, and more precisely to its lowest Ypresian stage (56 million–49 million years ago), locally named Sparnacian after the city of Épernay (Sparnacum in Latin). After hosting an estuary, the site became a tropical, swamp-covered delta with trees sticking out of its muddy waters. In a time interval of less than 5 million years, evolution ran its course, and a new array of fossils records the diversification of mammals, including the bones of a small, tarsier-like primate.

Today in the forest above the vineyard, there are a few quarries, since abandoned, dug into the Eocene mudrock—a water-logged, dark-colored clay with rust-colored streaks; it and the scattered clumps of primitive-looking horsetail plants give the impression that we have traveled back in time to this dawning age of mammals. In places, the boggy sediment is a true lignite coal that bears the imprint of 50-million-year-old leaves, twigs, and even entire tree trunks.

Although it lies above the tree line, this carbonaceous clay has a role to play in the vineyard. Winemakers have long recognized the properties of this material, which they call "ash" (cendres) or "black earth" (terre noire); they sprinkle it among the lower vines to amend the soil. Not only is it organic, but it is metal rich as well: it fights the iron deficiency typical of limy soil (ferric chlorosis), which weakens photosynthesis and makes leaves turn yellow. Moreover, the additive's dark color helps the soil soak up solar rays and heat the vines.

Getting back on the main road, our clockwise journey takes us next to the "nose" of the Reims Mountain, facing due east, before we veer south through the villages of Villers-Marmery and Trépail and cruise downslope—and back in time, across Cretaceous strata and a chalky soil well suited to Chardonnay. This eastern sector hence privileges the white grape.

When we reach Ambonnay and Bouzy on the southern flank of the Reims Mountain, the emphasis switches to Pinot Noir. Here, the Marne

River opens a cleft in the cuesta and carves the south-facing slope of the hill, where red wines have been produced throughout history and still are today.

THE MARNE VALLEY

The Marne valley penetrates the plateau westward, and isolates the Reims Mountain to the north from the Côte des Blancs to the south. The river flows through Épernay—co-capital of the Champagne wine industry, on a par with Reims. Then the valley narrows between Œuilly and Reuil before the Marne follows a twisting course along east–west fault-controlled segments that are separated by the river's short jumps to the north or south along perpendicular faults.

There is much more clay in the Cenozoic strata incised by the river. Soaking up rainwater, the slope is unstable, and vineyard lots are occasionally lost to slumping. The substitution of limestone by more recent marl and clay is caused by the regional dip of the strata, running slightly upward from west to east. The oldest layers, at the bottom of the pile, therefore intersect ground surface in the easternmost realm of the vineyard, around Reims, Épernay, and the chalky plains of Champagne. As we head west along the Marne, the beveled layers outcrop in successively younger stripes: Paleocene sandy limestone past Dormans, then Eocene clay beyond Château-Thierry.

There is still sufficient limestone around Épernay to favor Pinot Noir, but as we proceed into the valley, the transition to sand and clay gives an advantage to Pinot Meunier, better adapted to the clay-rich soils as well as to the cooler, more humid microclimate. The Meunier variety accounts for 80 percent of the vines in the Marne valley. As a result, winemakers who own lots both at the mouth of the valley, where Chardonnay and Pinot Noir are dominant, and downstream, where Pinot Meunier takes over, are fortunate enough to produce all three varieties and try out all blend combinations.

A good example is the Tarlant estate. Based in Œuilly, a dozen kilometers (about 7 miles) downstream from Épernay, the estate owns 14 hectares (35 acres) of vines distributed among four different townships. It makes a point of vinifying each grape and lot separately before blending, going so far as to promote special cuvées belonging to distinct geological strata—an unusual philosophy in Champagne, where it is more common to assemble a uniform, predictable product.

Showing us around the estate, Micheline Tarlant confirms that all fifty-five lots are vinified separately in stainless-steel vats or oak barrels, depending on the grape. At the end of winter, owners and cellar masters get together to taste wine from each lot and determine which ones to assemble and in what proportions to make the regular line of Champagnes for the year. They also decide on those three special blends that would best highlight the three different terroirs of the Marne valley.

The first "terroir Champagne" is labeled *Cuvée Louis*, named for family patriarch Louis Tarlant (1878–1960), and comes from a marine chalky limestone lot straddling the Cretaceous/Cenozoic boundary—a lot appropriately named Les Crayons (*craie* means "chalk"). Both Pinot Noir and Chardonnay are grown here, and are blended in equal proportions. Chardonnay raised on chalk brings mineral and citrus notes to the Champagne, although the blend with Pinot Noir results in a complex bouquet of honey, baked apple, and candied fruit.

The second terroir highlights the younger Eocene clay (Sparnacian stage) on the higher slopes of Œuilly. Named Cuvée Vigne d'Or, it relies solely on Pinot Meunier and gives off a bouquet of exotic fruit, with a touch of almond. Wine tasters detect lychee, passion fruit, mango, pineapple, hazelnut, and praline.

The third terroir is a sandy layer above the clay, marking the return of the sea over northern France and of a coastline running through Champagne—a terroir winemakers have nicknamed "rabbit sand" (*sable à lapins*), because it is now riddled with burrows. Chardonnay is the exclusive variety here, and yields a silkier Champagne than it does on chalk, with a bouquet of acacia and lime tree, butter and honey, brioche bread and hazelnut. Labeled *La Vigne d'Antan* (*antan* means "yesteryear"), it also happens to be the only vine lot that is ungrafted and planted directly into the ground, with no need for American rootstock. This miracle is possible only because of the sandy terroir: the phylloxera larva cannot cope with sandy soil, at least of this type and in this clime. The result: a rare opportunity to taste the otherwise lost authentic aromas of the original Chardonnay, hence the name Vigne d'Antan.

Above the sand layer, the next strata document the prehistoric sea's further advance across Champagne. Literally stuffed with seashells, they tell the story of a new Eocene stage, named the Lutetian (49 million–40 million years ago) because these layers are also found around Paris, which was named Lutetia in Roman times. In the Marne valley, Lutetian shelly sediments outcrop toward the top of the slope, at the level of the highest vineyards—namely at Fleury-la-Rivière, where the fossils are spectacular and beautifully preserved.

A TUNNEL THROUGH TIME

You can be a winemaker and also a rock collector and a geologist at heart. Such is the case with Patrice Legrand, who was groomed to take over his family's Champagne business in Vandières, on the Marne's right bank, 20 kilometers (12 miles) downstream of Épernay. Fascinated by the abundance of fossils in the area, he pored over the ground, becoming quite a connoisseur in paleontology, and decided with a group of friends to drill a tunnel into the hillslope to further explore the fossil grounds.

Together with his wife Anne, Legrand bought two houses in the nearby town of Fleury-la-Rivière, up a little valley, that had their backs to the hillslope and were flush with the fossil-bearing strata. Named *tuffeau de Damery*, the sandy limestone is 45 million years old, midway through the Lutetian stage, when the sea came as far east as Reims and Épernay, and was met along the low coastline by rivers draining off the midlands of eastern France. Fleury-la-Rivière is now located on the site of what was once a delta: a network of channels, separated by marshes of brackish water, with sandbars and stands of mangroves.

Legrand and his team set out digging a tunnel into the delta's sandy strata with a small pneumatic drill. Slowly but surely, they managed to pierce a 200-meter-long (660-foot) tunnel connecting one house cellar to the other. The idea was not only to extract a wealth of fossils from the excavation but also to leave some of the best specimens in place, still set in stone in the walls and ceiling of the gallery, carefully cleaned and artistically lit.

Open to the public by appointment,* La Cave aux Coquillages ("Cellar of the Seashells") is an exciting underground excursion into the lost

* La Cave aux Coquillages, 39, rue du Bourg-de-Vesle, Fleury-la-Rivière; phone: 03 26 58 36 43; website: www.lacaveauxcoquillages.fr/.

world of the Eocene. Combining technical achievement with a true sense of pedagogy and an artistic touch, Patrice Legrand and his team carved out niches along the tunnel to highlight the fossils discovered, as well as scenes of submarine life from Eocene times, such as crawling mollusks and swimming tropical fish frozen in action (figs. 10.4, 10.5).

The labyrinth of lagoons and shallow channels in the prehistoric delta revealed by La Cave aux Coquillages certainly discouraged most predators of the open sea, except for a few juvenile shark and skit. Consequently, the area was a protected sea-life nursery full of mud burrows and clumps of sea grass, where mollusks, goby fish, crabs, and other crustaceans crept along the seafloor.

The mollusk shells are especially beautiful; some have even kept their original colors. This remarkable state of preservation is due to a layer of green clay that covered up the shell-bearing layer and entombed the fossils, protecting them from chemical attack for over 40 million years. Among these fossils, the star species is *Campanile giganteum*, a mollusk hiding in a giant coiled seashell. It gets its name from the Campanile bell towers of Italy known for their stacked arcades, because its shell is likewise ribbed with multiple rings. The shells of this species measure up to half a meter (over a foot) long in the case of the oldest individuals, which

Figure 10.4 Fossils of the giant Campanile mollusk (*Campanile giganteum*), highlighted in the tunnel of Patrice Legrand's *Cave aux Coquillages*.

Figure 10.5 Scene of submarine life in the Champagne sea, 40 million years ago. (Model by Thierry Dupin, La Cave aux Coquillages.)

are thought to have lived for more than a century. *Campaniles* dwelled half buried in the bottom ooze to escape predators. At night they crawled around on their slimy bellies—in true gastropod fashion—in search of food.

The gallery of La Cave aux Coquillages, and the entire *tuffeau de Damery*, for that matter, is a spectacular *Campanile* cemetery. There are typically two or three giant shells per square meter, concentrated in a layer less than one meter (3 feet) thick. Above this layer, however, the sand is grayer and finer, with no more giant seashells. What led to their demise? A demographic explosion that caused food depletion, starvation, and population collapse? A change to the sea bottom in salinity, temperature, or water depth? The mystery remains.

As we exit the gallery, lost in Eocene thoughts, Patrice Legrand brings us back to the present with a glass of his excellent Champagne Legrand-Latour, fresh and lively with a fruity bouquet and a stream of fine bubbles—the evanescent souls of the old Lutetian seafloor.

CHAPTER ELEVEN

BORDEAUX

The vineyards of Bordeaux (fig. 11.1) are famous worldwide, and are second only to Languedoc as suppliers of French wine. They cover 100,000 hectares (250,000 acres) and produce 800 million bottles a year: wines for all tastes and budgets, from the regional and cheap simple Bordeaux appellation to the sublime Grands Crus and century-old vintages that command thousands of euros for a single bottle.

Bordeaux wines also come in all colors: even if the majority of the production is red wine, the region also offers rosés, dry whites—namely Graves and Entre-deux-Mers—and prestigious sweet whites, such as Barsac and Sauternes.

The Bordelais area offers a wide gamut of terroirs, ranging from limestone to marl and clay, as well as terraces of alluvial sand and gravel that constitute one of the most typical substrates of the vineyard. Additionally, its climate happens to be exceptional, since it is the only wine area in France—except for Muscadet at the mouth of the Loire valley—that enjoys a humid ocean climate.

Bordelais vines are planted a few dozen kilometers inland from the Atlantic coast, where two great rivers meet that spread gravel and cobbles over the land: the Garonne, running down from the Pyrenees, and the Dordogne, flowing out of the Massif Central. They converge downstream of Bordeaux to form a wide estuary known as La Gironde. Long before these recent rivers had deposited the gravelly layers of the terroir, sea-level fluctuations 40 million to 20 million years ago during the Eocene and

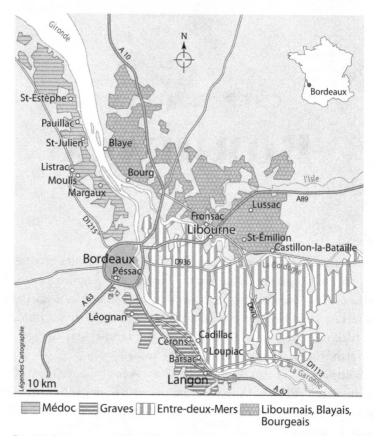

Figure 11.1 Bordeaux wine appellations. (Map by Legendes Cartographie.)

Oligocene periods caused the Atlantic to flood the lowlands many times over, laying down a substrate of limestone and marl (fig. 11.2).

Vine growing did not come naturally to the area. The Bordelais region lay in the western corner of the Roman Empire and suffered from its relatively cool and rainy ocean climate, but two factors made the wine industry take off. The first was the dynamic growth of the city of Bordeaux. Reacting against the high price of imported wine—caused by a monopoly of Roman merchants—the citizens of Bordeaux decided to plant their own vines. This emancipation was possible in part thanks to a grape variety well adapted to the cool climate of the Atlantic coast. Originally grown in the Basque country on the Spanish side of the Pyrenees, this *Biturica* grape made its debut in the Bordeaux area during the first century AD

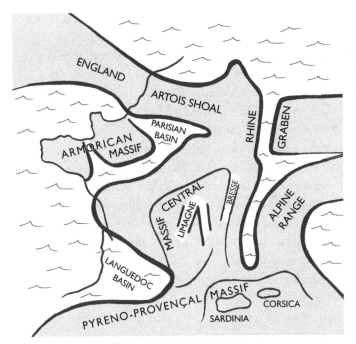

Figure 11.2 France during the Oligocene Epoch, 30 million years ago. The last marine transgressions flooded part of the Paris Basin, the Rhine rift basin, and the Aquitanian basin, where limestone and marl constitute the bedrock of the Bordeaux terroir. (Map by Pierre-Emmanuel Paulis.)

and is cited by Roman authors Pliny the Elder and Columella. It could well be Cabernet Franc or a close cousin, and by crossbreeding with other vines it ultimately yielded Cabernet-Sauvignon and Merlot, which did very well in the area, enjoying both climate and terroir.

The second factor contributing to the success of Bordeaux wines was the trade route provided by the Gironde estuary opening up on the Atlantic coastline. While the Mediterranean market was saturated and impenetrable, and the Paris market locked up by the nearby vineyards of Champagne, Burgundy, and the Loire valley, this connection to the Atlantic Ocean was a true blessing. All that Bordeaux wine needed was an opportunity to conquer the world market.

That opportunity came knocking in 1152 when French queen Eleanor of Aquitaine (1122–1204) annulled her marriage to Louis VII and wed Henry II Plantagenet, future king of England. This alliance opened up commerce on a grand scale between Aquitaine—the greater Bordeaux area—and the British Isles, starting with wine. Soon the traffic reached one thousand ships a year, loaded with barrels of light rosé wine called claret, a mix of black and white grapes.

Four centuries after Queen Eleanor, King Henry IV of France (1553–1610) took the Bordeaux vineyard one step further by calling in Dutch water engineers—the polder specialists—to drain swamps across France, particularly those of Médoc on the left bank of the Gironde estuary.

The Médoc area, destined to become one of the best terroirs in Bordeaux, is the spit of land between the Atlantic Ocean to the west and the wide Gironde estuary to the east (the name Medoc comes from the Latin *medio aquae*—"between waters"). On the Gironde side, the lowlands were spoiled by swamps and unfit for agriculture. Only after the Dutch drained the marshes could vine growing spread from the city of Bordeaux to the newly available land downstream, propelling the new branch of Bordeaux wines to stardom.

Yet the Dutch did much more than simply master the local waters. They were genuinely interested in the vineyards and bought land, often at low prices, in exchange for their work. In order to supply the Amsterdam market and their international clientele, their preference for hearty red wine over the British claret, along with sweet whites, launched Bordeaux winemaking in two directions that would make the southwestern vineyard world famous.

The Dutch were also conscious of the importance of terroir, experimenting with different soil types and testing their alliance with different grape varieties. The French "Gascon" winemakers of Aquitaine followed suit, beginning with the Pontac family that ran the Haut-Brion estate south of Bordeaux. In the 1650s, Arnaud III de Pontac was one of the inventors and advocates of the notion of "Cru" to define a distinct terroir characterized by its soil and microclimate. He launched the production of full-bodied red wines that aged and traveled well, and set the standard for all great Bordeaux reds to come. This "New French Claret," as it was called, was a big success. Both Louis XIV in France and Charles II in England served Haut-Brion at their royal tables.

While great vineyards were spreading on the left bank of the Gironde estuary, the right bank also underwent considerable development: the steep slopes above the harbors of Blaye and Bourg produced fine wines that later took on the official appellations of Côtes-de-Blaye and Côtes-de-Bourg. In our clockwise survey of the vast Bordelais territory, we will begin with these right-bank townships (located at "noon" on our map), since they happen to reveal the oldest rocks in the area.

THE RIGHT BANK: CÔTES-DE-BLAYE AND CÔTES-DE-BOURG

The oldest substrate of the Bordelais region is not that old—it is actually the youngest of all substrates underlying French vineyards. The sedimentary strata record the oscillations of the Atlantic shoreline as the sea level rose and sank repeatedly during the Cenozoic, between about 40 million and 20 million years ago.

The earliest, lowermost strata pick up the geological history of France where we left off in Champagne, with mammal species fast evolving, and take us through the next period of the Cenozoic, named the Oligocene (34 million to 23 million years ago). The overall tendency at the time was a rise in sea level, a rougher climate, and a bout of natural selection, pruning the evolutionary tree both on land and in the sea.

To leaf through this new chapter of geological history, we follow an itinerary up the right bank of the Gironde estuary, starting at Blaye harbor. Fifty kilometers (30 miles) downstream of Bordeaux, Blaye (pronounced "Bly") has a long history. In his *Commentaries on the Gallic War* that relate his expeditions in France from 58 to 52 BC, Julius Caesar mentions a Gallic stronghold named Blavia Santorum that might well be the original citadel of Blaye. The Romans built their own fortress on the site, and for good reason: it is one of the rare rocky spurs overlooking the estuary, facing sandbars in the middle of the current that further help control the passageway—not only navigation up and down the Gironde, but also the river crossing east to west, leading to the left bank and to the land road upstream to Bordeaux. For the same reason, Blaye would later become a major gateway on the well-traveled pilgrimage route to Santiago de Compostela in Spain.

Much later, Louis XIV had an artillery fort built on the site, masterminded by his military architect Vauban, with ramparts following the rim of the rocky platform and massive towers at the corners.

Blaye ended up losing its military importance after a final confrontation with the English fleet in 1814 during the Napoleonic wars. The citadel's geostrategic advantage was sacrificed when the first stone bridge was

built in Bordeaux across the Garonne River, but it hung on to its prominent position as a wine-producing and wine-shipping district. Since the
Middle Ages, the hills and plateaus stretching along the estuary and inland from the city have never stopped producing wine.

Today, nearly 7,000 hectares (17,500 acres) of vines qualify for the appellations Blaye, Côtes-de-Blaye, and Premières-Côtes-de-Blaye, and
produce mostly red wines, made mainly from Merlot and Cabernet-
Sauvignon—the two starring grapes of Bordeaux—with occasionally a
minor contribution from Cabernet Franc and Malbec.

The terroir of Côtes-de-Blaye is remarkably diverse, well justifying the
coexistence of several grape varieties. Underpinning the vineyard is the
basal limestone, pretty much hidden from view in most places, overlain
by clay-rich strata, sand layers, gravel beds, and even a recent sprinkling of
wind-driven loess, as young as the most recent Ice Ages.

The oldest limestone outcrops in the cliff face under the citadel. It was
brought up to the surface by an anticline fold. This Blaye limestone dates
to the Lutetian age of the Mid-Eocene, approximately 45 million years
ago. It indicates the presence of a shallow sea at the time, but a sea that was
progressively retreating westward. The shell-packed sediment contains a
few surprises, namely rib bones that belonged to Sirenian sea mammals—
ancestors of today's manatees and sea cows.

Above the Blaye limestone lie oyster-rich clays that indicate shallower
receding waters, and that are often exposed alongside vine lots where
plowing has occurred. Above this clay we reach the layers that are directly
involved in the terroir, starting with the Plassac limestone. It is a lacustrine sediment, proof that the sea had fully withdrawn from the area at
the time, replaced by a lake. But the seashore lay close by, somewhere in
the present-day Médoc area on the other side of the present-day estuary.

Climbing higher up in the Blaye vineyard, we reach another layer of
oyster clay, signaling the return of the sea, first in the form of brackish lagoons, then as a frankly marine limestone that marks a further deepening
of the waters, which have resumed their inland incursion. Present on the
hilltops, this cap rock is known as Saint-Estèphe limestone, since it also
outcrops in the left bank around the town of Saint-Estèphe.

This limestone and clay sequence is particularly well suited to Merlot, which is the dominant grape variety in Côtes-de-Blaye; it gives the
wine its fruity character. Farther inland, where patches of recent sand and

gravel overlie the Eocene sediments, the soil has greater drainage potential and a warmer temperature profile. It supports other grape varieties, such as Cabernet-Sauvignon and Malbec.

One of the characteristics of Bordeaux wines, and of Côtes-de-Blaye among others, is the blending of several grape varieties, also called *assemblage*, in order to bring together their complementary qualities. The flip side of the coin, for a geologist, is that it is more difficult to pick out the contribution of terroir in a blend; whereas a pure varietal, as we saw in Alsace, Burgundy, and the Loire valley, is more telling in this respect.

Some of the Côtes-de-Blaye wines can be made solely from Merlot. More often, each will be a blend of approximately three-quarters Merlot to one-quarter Cabernet-Sauvignon, with sometimes a touch of Cabernet Franc or Malbec. Merlot brings to the wines a fruity character with soft, voluptuous tannin, whereas Cabernet-Sauvignon delivers body and good aging potential.

Getting back on the road, we head south along the Gironde estuary toward the upstream harbor of Bourg, at the confluence of the Dordogne and Garonne Rivers. We still notice limestone strata along our journey, but since we are moving away from the bulging Blaye anticline the strata dip downward, with the lowest and oldest ones disappearing underground and being replaced at our level by younger ones, higher up in the sequence and spared here from erosion.

Along the way, we cross the time boundary between the Eocene and Oligocene periods, dated at 34 million years, and enter a time when the sea advanced yet once more over the land toward Monbazillac and Bergerac, and coastal marshes spread as far east as the city of Agen.

In warm, shallow waters, many shells accumulated on the seafloor, together with echinoderm fragments—animals of the sea urchin family—to form a hard, calcareous sediment known locally as Bourg limestone and more formally as Asteria limestone (*calcaire à astéries*) in light of the numerous starfish fragments found in the rock. *Calcaire à astéries* is the main building stone of the region: easy to carve and aesthetically pleasing with its warm, golden hue, it makes up most the stone façades of Bordeaux and other cities of Aquitaine.

The limestone cliffs around Bourg-sur-Gironde are riddled with quarries and natural caves, most notably at Prignac-et-Marcamps up the valley of the Moron River, and some galleries have been transformed into mushroom farms. Our Stone Age ancestors were well aware of the lodging benefits provided by limestone caves: a prehistoric habitat was discovered in 1881 by François Daleau, a Prignac winemaker who had developed a keen interest in prehistoric research and digs.

The little grotto Daleau uncovered was located in a limestone knoll in the middle of his vineyard. It contained thousands of small ivory and flint tools and jewelry, and its walls were decorated with pictures of horses, mammoths, aurochs, and deer carved into the rock; these were tentatively assigned to the Aurignacian culture from more than thirty thousand years ago. The grotto was probably occupied earlier yet by our Neanderthal cousins, to whom paleontologists credit a three-holed flute sculpted out of a hollow vulture bone discovered on the site.

Besides the limestone buttes of prehistoric repute, the Côtes-de-Bourg vineyard also covers a sandy terroir farther inland toward Pugnac and Saint-Trojan—sands of a vast delta that met the sea at this location. The fossilized delta is now exposed by erosion as a gently rolling landscape.

The Côtes-de-Bourg vineyard therefore enjoys the same kind of terroir diversity as Côtes-de-Blaye. Geologists have identified in the appellation zone no fewer than twenty different soil types, ranging from limestone-derived calcosols and rendosols to gravel peyrosols, clay-rich brunisols, and delta-based luvisols.

The range of different soils and grape varieties makes for a vast array of wines in Côtes-de-Bourg, especially since the area counts no less than five hundred wine producers. If we were to pick only one, Château Mille Secousses is especially representative of the history and flavor of the place.

The estate's vines are planted on a clayey limestone soil facing south toward the Dordogne River. The higher slopes deliver a fancy Côtes-de-Bourg, and the lower slopes simply qualify for a Bordeaux Supérieur appellation, but yield one of the best in its category. Oenologists credit the wines' supple tannins and finesse to the quality of the terroir, with the Côtes-de-Bourg in particular evolving over time toward a very floral bouquet, with notes of truffle and forest undergrowth.

The estate is deeply steeped in history. It was founded in 1632 by Dutch merchant Jean de Ridder, who drained marshes along the Dordogne River, a few kilometers upstream of Bourg, and planted vines on the reclaimed land to supply the Amsterdam wine market. The entrepreneur gave a final touch to his estate by landscaping his gardens as stylish

Versailles-like *jardins à la française*, complete with fountains and canals. The stunning ensemble caught the attention of the French royal court when it summered at Bourg in 1650. Governing Bishop Mazarin, Queen Mother Ann of Austria, and the twelve-year-old future King Louis XIV very much enjoyed taking boat rides to Jean de Ridder's estate, conveniently equipped with a dock, and his hunting pavilion quickly reached an enviable notoriety.

How the estate became known as Mille Secousses ("A Thousand Jolts") is an amusing side story. Since a major fault runs along the estuary, could it be that winemakers had felt jolts in the past? The real explanation is much more . . . down to earth. While the queen, future king, and bishop enjoyed a smooth sail to the estate, the rest of the royal court had to take horse-drawn coaches along a particularly bumpy dirt road. Mademoiselle de Montpensier, an older cousin of Louis XIV, joked about the journey by naming the Dutch mansion Château of a Thousand Jolts (Château des Mille Secousses), and the name stuck.

SAINT-ÉMILION

As we follow the Dordogne River via route D 670 toward the southeast and the cities of Fronsac, Libourne, and Saint-Émilion, we notice a greater proportion of clay, corresponding to the transition between the Eocene and Oligocene periods, 35 million years ago. Back then, a drop in sea level uncovered the limestone seafloor and transformed the site into a floodplain; rivers deposited green-and-white clay laced with meandering, sandy channels. The fine-grained clay, known as Fronsac molasse, is a blessing for terra-cotta tile manufacturers as well as for winemakers—the argillaceous hillslopes are particularly well suited to Merlot. This terroir on the outskirts of Libourne is home to the plumy, powerful, and beautifully structured Fronsac and Canon-Fronsac appellations that blend Merlot with a touch of Cabernet Franc and Cabernet-Sauvignon: they boast spicy and mineral flavors, and age remarkably well.

The City of Libourne was founded in 1268 by order of English king Edward I, and named for the king's faithful lieutenant, Sir Roger de Leybourne (1215–1271), who supervised the enterprise. The idea was to build a harbor at the confluence of the Dordogne and Isle Rivers that would compete with Bordeaux and attract part of the flourishing wine trade.

This historic connection with the British Isles stayed very much alive over the centuries, and the Libournais wines were always held in high esteem across the Channel. In contrast, it took the construction of the French railroad in 1853 for these wines to achieve recognition on the French market, and yet another century before they reached international fame, after World War II. In the past few decades, the star appellations of the region—Saint-Émilion and Pomerol—have made up for their slow start by becoming two of the most famous and expensive wines in the world.

Ironically, Libourne's vineyards had been there almost two millennia, cultivated on a rich limestone plateau between the Isle and Dordogne Rivers. We know how old the vineyards are thanks to Gallo-Roman mosaics picturing harvest scenes that were uncovered at the foot of a hill on the estate of La Gaffelière (a Saint-Émilion Premier Cru Classé), and because Latin poet Ausone (AD 309–395), born in Bordeaux, praised the local wine. Ausone gave his name to another Grand Cru of Saint-Émilion, Château Ausone, an estate purportedly established on the ruins of his Roman villa.

Saint-Émilion is one of the most picturesque villages of the Bordeaux area, perched on a limestone butte with a steep slope plunging down to the Dordogne valley to the south and a shallower slope leading to the Barbanne stream to the north. It owes its beauty to the handsome, readily available limestone (*calcaire à astéries*, locally known as *calcaire de Saint-Émilion*) and to the building talent of its Benedictine and Augustine monks, who settled on the site where evangelist Saint-Émilion retired as a hermit in the eighth century.

Over the next three centuries, the monks transformed the hermit's cave into a monumental underground chapel, a masterpiece known as the Monolithic Church; its entrance is flush with Saint-Émilion's marketplace (place du Clocher) in the lower part of town. Ramparts were also built around the town, and a Romanesque church was erected above the monolithic underground chapel. So were handsome medieval houses, all featuring the same golden limestone. Walking the streets of Saint-Émilion, you will notice crisscrossing layers in some of the building blocks: the mark of a strong current that raked the sea bottom some 30 million years ago, when the limestone was formed.

Viewed from the ramparts, beyond the glistening terra-cotta roofs, the vineyards of Saint-Émilion roll out to the horizon, straddling a number of different terroirs that make this prestigious appellation a highly complex one, where one wine can be sublime and the next one ordinary and overpriced.

The official classification of Saint-Émilion wines is thoroughly revised every ten years—a very modern approach, contrary to the Graves and Haut-Médoc classification (around the city of Bordeaux) that was established in 1855 and remains frozen in time. Saint-Émilion's dynamic classification comes with its share of local politics, favoritism, and legal battles. The 2006 classification, for instance, was challenged by demoted estates; it was overturned, and the plaintiffs were reinstated. The classification's long list grants the distinction of Grand Cru to about two hundred estates, in recognition of the higher quality and good aging potential of the wine over "ordinary" Saint-Émilion. The short list, moving up the ladder, sets apart about fifty Grands Crus Classés—of greater quality yet—and features at the top a final tier of fifteen or so Premiers Grands Crus Classés, culminating with the two very best estates: Château Ausone and Château Cheval Blanc.

To explore Saint-Émilion and its many vineyards, we follow route D 670 across the Dordogne alluvial plain between Libourne and Castillon-la-Bataille. The road hugs the foot of the limestone plateau, scalloped by numerous gullies. Splitting off route D 670, a side road (D 122) follows one of the gullies up to Saint-Émilion, past the picturesque, vine-planted terraces of Clos la Madeleine (still awaiting Grand Cru recognition). The château on the top terrace is built flush against the limestone strata that cap the Saint-Émilion sedimentary sequence; but one of the best places to view the sequence in cross section lies 5 kilometers (3 miles) farther east, along the vineyards of Château Mangot, past the village of Saint-Étienne-de-Lisse (via route D 243). As we drive around a bend overlooking sloping vines to the right, and just before reaching a pine grove, an embankment runs off from the road along the edge of the vineyard (fig. 11.3).

At the foot of the embankment, the vines extend their roots into Fronsac molasse—the floodplain clay that we encountered in the Fronsac appellation. This is the lowest stratum making up the Saint-Émilion terroir, dating, as we saw, to the Eocene/Oligocene boundary, 35 million years ago. Merlot is at home here, and accounts for 85 percent of the blend assembled by Château Mangot, spiked with 9 percent of Cabernet Franc

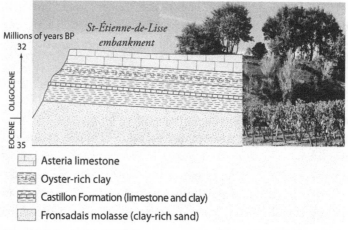

Asteria limestone

Oyster-rich clay

Castillon Formation (limestone and clay)

Fronsadais molasse (clay-rich sand)

Figure 11.3 On the edge of the Saint-Émilion plateau, an embankment displays the rocky strata underlying the vineyard: limestone, clay, and clayey sand. *BP* signifies before present.

and 6 percent of Cabernet-Sauvignon, although proportions vary. Recognized as a Grand Cru, Château Mangot has an earthy bouquet, with notes of licorice, menthol, and cooked prune, and lingering, fresh mineral flavors.

As we look upslope at the embankment above the vines, its lower half consists of several feet of green clay encasing a fine bench of pinkish limestone, known as the Castillon Formation—evidence of a temporary lake that covered the area 34 million years ago, after the alluvial plain episode. The lake progressively became a swamp before drying up completely. Above the formation lies a darker layer of clay, full of oyster shells, indicating the return of a shallow sea. Finally, the water got deeper, and a coastal sequence of urchin and starfish debris collected on the sea bottom to form the famous Saint-Émilion limestone (*calcaire à astéries*) that caps the embankment and makes up the hard floor of the plateau.

This complex cycle of marine transgressions and regressions spans a time interval of approximately 5 million years, 35 million to 30 million years ago. Today, this multilayered sequence provides a healthy substrate to the vineyard. The slightly porous top limestone readily absorbs rainwater, while the basal Castillon clay forms an impermeable seal that stops the downward infiltrations, storing dampness at depth and guiding any excess water underground until it intersects a hillslope and trickles out as a spring. This latter sequence happens in the very town of Saint-Émilion, where water disgorges from the rock face (Source de la Médaille) and at the bottom of a well (Puits des Jacobins).

Most Saint-Émilion Grands Crus are blends that mix grapes from the limestone plateau with those from the clay-rich slopes. A good example of this double terroir can be found at Clos Fourtet (fig. 11.4), a Premier Grand Cru Classé whose vines start on the limestone plateau, opposite the western gate to the town and the Romanesque church. This was the site of Camp Fourtet, a military encampment that defended the town entrance during the Middle Ages. Today, the estate and its château are surrounded by a low stone wall (hence the name Clos).

Within these walls, the Clos Fourtet vineyard stretches over the edge of the limestone plateau and cascades down to the west in a series of flat terraces and connecting slopes—the terraces betraying hard limestone strata and the sloping ramps covering softer layers of clay that creep and spread. On this estate, Merlot accounts for approximately 85 percent of the blend, Cabernet-Sauvignon for 10 percent, and Cabernet Franc for 5 percent.

Besides limestone and clay-rich molasse, there is a third type of terroir at Saint-Émilion, rather atypical and very precious, that makes up the

Figure 11.4 The vineyard of Clos Fourtet, one of the fifteen Premiers Crus of Saint-Émilion AOC, faces the Collegiate Church (*left*), with the spire crowning the Monolithic Church in the background to the right.

northwestern corner of the appellation. It consists of a terrace of gravel and small cobbles deposited by the meandering courses of the Isle River and of its confluent, La Dronne, during the recent Ice Ages. This alluvial terrace is rich in rounded pebbles of black flint and golden quartz, rolled down from the Massif Central. Tens of centimeters to over several meters thick, the gravelly blanket lies over the much older sandy clay of the Fronsadais molasse.

Low in nutrients but providing good drainage, the terrace suits Merlot but even more so Cabernet—a slow-maturing variety that benefits from the solar heat stored and radiated back by the pebbles. The aromatic palette of the grapes (be they Merlot or Cabernet) raised on such alluvial terraces is known to become superbly complex over the years—so much so, in fact, that two of the top Premiers Crus of Saint-Émilion belong to this gravelly terroir: Château Figeac and Château Cheval Blanc.

A world-famous wine, Château Cheval Blanc brings together the opulent, fleshy body of wine raised on clay and the tannic finesse brought by the gravel component. Each lot is vinified separately to bring out the character of the terroir, and only then are the components blended: a majority of Cabernet Franc (around 58 percent), with Cabernet-Sauvignon and Merlot playing second fiddle, which is rare in the Libournais area. The bouquet leans toward forest undergrowth, and the wine is long on the palate with silky tannins, courtesy of the gravel beds.

Château Figeac also owes its great class and reputation to the gravelly terroir. The estate, less than a kilometer (six-tenths of a mile) from Cheval Blanc, is founded on the site of a second-century-AD Gallo-Roman villa that belonged to one Figeacus. You can still run across stone ducts on the estate that date to antiquity.

At Château Figeac, the vineyard covers three low hills, called *croupes* (literally, "rumps"), that are blanketed by a layer of pebbles 5 to 10 meters (15 to 30 feet) thick; these are kown as fire gravel (*graves de feu*) because of their fiery hue in the sunlight. There is no lack of sun here, to the point where one of the mounds is called Croupe de l'Enfer ("Hell Hill").

Again, it is Cabernet that reigns supreme on such gravelly terroir, with its Cabernet Franc (35 percent) and Cabernet-Sauvignon (35 percent) ruling over Merlot (30 percent). This gravel-induced Cabernet exception extends northwest to the outskirts of Libourne, crossing a neighboring appellation that has risen to stardom: Pomerol (see box).

POMEROL

Pomerol is a small appellation covering 800 hectares (2,000 acres) of vines northeast of Libourne. It is situated on a clay-rich plateau covered by pebbles and cobbles, separated from the vast Saint-Émilion plateau by a tiny river. The subtle difference in soil and microclimate is sufficient, however, to make Pomerol a very special place, yielding a very special wine.

The name Pomerol probably derives from the Latin word *poma*, which stands for "seedy fruit" and grapes in particular. Vines were grown here as early as the first century BC, but it wasn't until the eighteenth century that the wines reached some level of notoriety, starting with the whites. Then the reds gained momentum, produced from Cabernet Franc and Malbec, but still failed to make it into the upper cut of Bordeaux Grands Crus. As a result, there are no châteaux in Pomerol, only modest farmhouses. It was the introduction of Merlot at the turn of the twentieth century that propelled the vineyard skyward, and by the 1960s made Pomerol one of the most sought-after and expensive wines in the world (600 euros for a bottle of recent "bottom-price" Petrus, and 3,000 euros for a ten-year-old good vintage from the estate). This success story results from the perfect match between Merlot (90 percent of the vineyard) and a unique terroir of thick clay covered by a blanket of gravel.

Like the westernmost section of Saint-Émilion, Pomerol's bedrock is covered by alluvial deposits of sand, gravel, and cobbles, spread across the land by the meandering of the Isle and Dordogne Rivers as they migrated west over time, digging deeper on the way and leaving behind them terraces that become progressively younger in stepwise fashion. On the westernmost edge of Pomerol, along route D 1089, the permeable blanket of recent sand and small ochre-colored gravel yields a supple wine, fruity and delicate, with light tannins.

Traveling eastward up the plateau, we reach the highest and oldest alluvial terrace, aged about one million years, when the rivers rushing out of the Massif Central experienced high discharge rates and dropped coarse gravel and cobbles. The layer is less than 2 meters (6 feet) thick and covers much older sand and clay molasse dating to the Oligocene alluvial plains, 30 million years ago. It is this rare association of coarse gravel over fine clay that creates the exceptional terroir of the top-of-the-line Pomerol Grands Crus, especially where the clay comes very close to the surface, as it does in the renowned Petrus estate.

The thin gravel cover provides drainage and accumulates solar heat at the foot of the vines. At a depth of a meter (3 feet) or so, the roots reach the clay and extend deep in search of water—a downward extension facilitated by the clay's contraction during dry periods which opens up desiccation cracks. The augmented contact surface between the root system and the iron-rich clay charges the Merlot grape with exceptional characteristics. Pomerol raised on clay is purple with bluish streaks, gives off an initial fruity bouquet of blackberry, black currant, and raspberry, and is graced with silky, velvety tannins. Over time, the terroir-derived chemistry truly kicks in: aromas become powerful, animal-like, with notes of truffle, roasted coffee, and cigar-box tobacco.

Before leaving Saint-Émilion, it is worth visiting the tail end of the appellation to the west, nestled in a meander of the Dordogne. This area is covered by recent gravel and sand from the river, and for that reason it used to be called Sables-Saint-Émilion ("Sand-Saint-Émilion")—a rare case of terroir type making it into a wine name—but the distinction did not last. Today, wines from this area are simply called Saint-Émilion, and some even reach Grand Cru status when they meet the proper standards.

Such is the case of Château Gueyrosse on the banks of the Dordogne, where Yves Delol and his daughter Samuelle produce silky, expressive wines that stand up to their more famous cousins of the central slopes and plateau of the appellation. Here the vineyard slopes down nearly to the waterfront, separated from it only by the château and its landscaped park. Although this is gravelly terroir, the winemakers stick to Saint-Émilion's classic proportions of grape varieties, with Merlot dominant (85 percent) and Cabernet Franc (12 percent) and Cabernet-Sauvignon (3 percent) playing minor roles. The fermentation yeast is indigenous, with no artificial spiking; and because the wine is matured in oak barrels that have already been used for three or four years, the oak tannin has mellowed out and doesn't overwhelm the delicate balance of natural tannin in the wine. All in all, despite its off-center location near the river, this is a classic Saint-Émilion, true to tradition, that qualifies most years for Grand Cru status.

ENTRE-DEUX-MERS

The Dordogne River, which runs roughly east to west from the Massif Central to the ocean, and the Garonne River, which joins it from the southeast, frame a vast, triangular area known as Entre-deux-Mers ("Between Two Seas"), a rather exaggerated name for two rivers, however wide.

Carved up by small confluents, this vast vine-growing region is a rolling landscape of low hills and flat-bottom valleys, its 7,000 hectares (17,500 acres) of vine interspersed here and there with orchards and forest-covered crest lines.

The underlying bedrock is once again *calcaire à astéries*. Called Bourg limestone and Saint-Émilion limestone farther north, it is known here as Entre-deux-Mers limestone, but it is the same thick sequence of seashell debris laid down by the marine transgression that penetrated the Bordelais area some 30 million years ago.

For geologists, the extra appeal of Entre-deux-Mers is that it records the subsequent episode of the region's geological history in the higher reaches of the vineyard spared by erosion. This later sequence consists of lacustrine limestone, interspersed with clay-rich molasse—an indication that the inland sea once again started retreating westward, replaced by lakes, marshes, and floodplains, around 25 million years ago.

As we drive to the southern edge of the Entre-deux-Mers plateau, overlooking the Garonne River, we reach the uppermost layer of the pile. Above the floodplain deposits lies a final cap rock of marine limestone, recording the last high stand of the sea in Aquitaine—an episode ushering in the Miocene epoch and logically called the Aquitanian stage.

At Sainte-Croix-du-Mont, a footpath follows the cliff ledge below the castle and church and reveals a spectacular layer of oyster shells several meters thick, testimony to this final marine transgression. The pristine shells look as though they were dumped there overnight after some giant feast, but they are aged nonetheless some 20 million years.

Entre-deux-Mers wine is fittingly a perfect match for oysters and seafood in general. A blend of Sauvignon and Sémillon grapes, it is dry and elegant, with a bouquet of citrus and passion fruit. At the southern edge of the plateau overlooking the Garonne River, there is an interesting shift in microclimate and in the type of wine produced. In the fall, morning fogs rise from the river and roll over the vineyards, triggering the spread of various molds on the grapes; these pump out water from the pulp and concentrate the sugar. The result is a naturally sweet wine that is produced in three adjacent townships on the right bank of the Garonne: Sainte-Croix-du-Mont, Loupiac, and Cadillac. But the most famous and precious sweet white wines of Bordeaux are produced across the river on the left bank: the world-famous Barsac and Sauternes.

SAUTERNES: THE ULTIMATE SWEET WHITE

On the left bank of the Garonne, 40 kilometers (25 miles) south of Bordeaux, the vineyards of Barsac and Sauternes are caressed by the autumn mold-producing fogs that rise from the main river and also from the

Ciron, a small, lazy-flowing confluent marking the boundary between the two appellations (Barsac to the north, Sauternes to the south).

The unique biochemistry that gives rise to the precious sweet wine requires a delicate balance. Morning fogs favor the development of *Botrytis cinerea*, a microscopic fungus that grows on the grape skin and inside the grape. The parasite feeds off the water and acids contained in the pulp, and pierces tiny openings in the grape skin. The Sémillon grape is particularly sensitive to the process, because it has a very thin skin.

If the weather remains humid throughout the day, *Botrytis cinerea* can get out of control, and other opportunistic fungi will join it. At this stage, known as gray mold (*pourriture grise*), the grapes are ruined and the harvest is lost. For the magic to occur, the morning fog needs to dissipate by early afternoon and make way for sunshine. Consequently, the growth of the fungus is limited to just the right amount, and the solar heat also drives the evaporation of pulp water inside the grape through the microscopic openings in its skin, helping to concentrate sugar and aromatic molecules. The success of the sequence hinges on the timely disappearance of the fog, and this is where the local topography comes into play. The slopes of the vine-growing hills—low-relief *croupes*—drive the flow of cold, humid air downward through sheer gravity, and out of the vineyards by late morning.

When the fungus growth is finely tuned in this manner by the microclimate, winemakers refer to it as "noble rot" (*pourriture noble*). Brownish spots break out on the skin of the grape, which proceeds to shrivel and shrink. The grape is then said to be "sun-baked" (*rôti*) (fig. 11.5) or candied (*confit*). Its acidity has dropped, transformed by fungal metabolism into glycerol molecules, which provide the wine-to-be with sugar, smoothness, and an aroma of candied raisin.

When they reach this stage, the Sauvignon and Sémillon grapes are harvested. Not only are they hand picked, but one should even say "finger picked," because only the most candied grapes of each bunch are collected. The others are left to ripen until the next inspection. Four to five rounds of picking take place over a harvest campaign lasting close to a month. Through this rigorous selection process, one vine plant ends up yielding a single glass of Sauternes, whereas in other vineyards the rule of thumb is more like a full bottle of wine per plant. Add to this the fact that in some bad-climate years an estate will not even allocate the grapes to

Figure 11.5 In the Sauternes area, only the ripest "sun-baked" grapes are picked; enriched with sugar and glycerol by microscopic fungi known as "noble rot."

Sauternes production (downgrading to cheaper wine), and you will understand why this special wine is so expensive.

The interaction of climate, grape, and fungus is so important that you might expect terroir to play only a minor role at Sauternes, but this is far from true. In some places, winemakers have gone out of their way to improve their land, as was the case for the most famous estate of Sauternes: Château d'Yquem.

The estate's name is probably derived from the medieval word *Ayquem*, meaning "bearing a helmet." Château d'Yquem has been known as a winemaking estate since at least the sixteenth century and the reign of French king Henry IV. Already at that time, local archives mention late harvests, although the grapes were probably naturally sweet, without the need for noble rot and multiple harvest rounds.

As was the case farther downstream in Médoc, Dutch engineers were called in to drain the wetlands and improve runoff down the slopes of clay-rich hummocks that were otherwise well suited to vine growing. This human intervention was key in ensuring the remarkable destiny of Château d'Yquem: the 112-hectare (280-acre) estate is crisscrossed by a drainage network that reaches a combined length of 100 kilometers (60 miles).

Without this sophisticated drainage system, the clay dome of Château
d'Yquem would never have hosted such a successful vineyard.

Besides the well-controlled drainage, the terroir also benefits, as it does
in Pomerol, from superficial layers of gravel laid down as alluvial terraces
by the Garonne River as it carved its way down to its present location during the recent Ice Ages.

Château d'Yquem's clay dome is covered by one of the oldest, highest-
perched terraces, probably close to one million years old. Classified as a
"Type 2" terrace by geologists, it is characterized by fine gravel coated in
clay, with small cobbles of white and golden quartz. Their natural drainage capacity and the sun-derived warming they provide the vine stock
contribute to the estate's success. As for the grape varieties involved, the
Château d'Yquem blend typically consists of 80 percent Sémillon and
20 percent Sauvignon.

A few hundred meters downslope along road D8, which leads to the
Garonne River, a slightly less famous Premier Cru of Sauternes is established on a lower gravel terrace. Named Château Suduiraut, it, too, boasts
an elegant manor, surrounded in this case by geometric gardens *à la française* designed by Le Nôtre, the same royal gardener who drew up the gardens of Versailles for Louis XIV. Here the gravel terrace is classified as a
"Type 3": slightly younger than the terrace of Château d'Yquem, probably
around five hundred thousand years old, its gravel is mixed with less clay
and more sand, and contains pebbles of white quartz as well as green volcanic cobbles rolled down from the Pyrenees Mountains. Here as well,
the Sémillon grape reigns supreme (90 percent of the blend), mixed with
a small fraction of Sauvignon (10 percent).

To enjoy a good Sauternes, however, there is no need to limit oneself to
expensive, big-name estates and expensive Premiers Crus. A case in point
is the small estate of Cru Arche-Pugneau. Its owner, Francis Daney, owns
a few lots next to the estates of Château d'Yquem and Château Suduiraut
that enjoy the same mix of sand and gravel over clay. Besides his classic
cuvées, which hold up rather well with respect to those of his illustrious
neighbors, Daney loves to try out new ideas. He elaborated an original
blend that he labeled *L'Intemporel* ("Timeless"): every year, he assembles
half a vat's volume of the new vintage with progressively smaller fractions
of previous vintages going back about ten years, with each contributing
their distinct aroma and personality to the blend—a celebration of wine
and time flowing by.

MÉDOC AND MARGAUX

In our clockwise tour of the Bordeaux wine region, we have now passed six o'clock—the southernmost stop on our journey—and head back north, now following the Garonne's left bank. This bank is covered with old gravel from the river, and the winemaking area has logically taken on the name of Graves, up to the city of Bordeaux. The rock substrate at depth might well change slightly along the way, but it is predominantly the gravel layer that sets the tone for the wine. This happens to be the only place in France where the name of a wine appellation—Graves—concerns the nature of the soil. Gravel here takes on many colors, and the finer sand coated with iron oxide gives the soil a predominantly brownish hue.

The southern part of the Graves area produces sweet whites, as we just saw—the starring appellations of Sauternes and Barsac—but as we drive north, the heartland of Graves produces reds as well. These are interesting reds, since we have reached the midway point in terroir and climate, where Cabernet-Sauvignon reaches 50 percent of the production, on a par with Merlot. Graves red wines achieve a smooth balance, with a marked bouquet of black currant, and their best vintages age remarkably well. As for the Graves dry whites, long the main wine type produced in the region, they rank as the best dry whites of Bordeaux and are made, like their sweet cousins of Sauternes, from Sémillon and Sauvignon grapes.

The quality of Graves wine peaks as we near the city of Bordeaux, a fact recognized in the form of a separate appellation awarded in 1987: Pessac-Léognan. Here the gravel spread by the past meandering-about of the Garonne River is particularly spectacular: white, blond, pink, and red quartz, jasper and agate, green volcanics and black chert. Vine growing here also goes back to antiquity, and while the rapid growth of the city of Bordeaux has taken over much of the agricultural land, a few oases of vines proudly survive. They're surrounded by roads and buildings, and put out some of the best wines in the world. The most famous of these "suburban" estates is Château Haut-Brion, ranked among the top five Bordeaux Premiers Crus in the official 1855 classification; it still sits at the top of the wine charts today.

The good match between gravel and Cabernet-Sauvignon—the grape variety needing the heat from the sun-warmed stones to mature properly—long recognized in the area, is such that for the first time in our tour of the Bordeaux region, Cabernet-Sauvignon ranks above Merlot in the blend proportions—49 percent to 46 percent—with a 5 percent balance for Cabernet Franc and Petit Verdot. The leading role of Cabernet-Sauvignon intensifies as we drive north past Bordeaux along the vast land spit of Médoc, framed by the Atlantic Ocean on one side and the Gironde estuary on the other.

Médoc owes its name to the Latin *medio aquae*—"between two waters"—and is covered by pine trees on the seaside (the Landes forest) and vineyards on the riverside. Because of marshy conditions, it wasn't until the seventeenth century and the drainage by Dutch engineers, as already mentioned, that vine growing spread northward into the peninsula. Since then, the rise of Médoc wines has been spectacular, with 13,000 hectares (32,500 acres) of vines bearing such famous names as Margaux, Saint-Julien, Pauillac, and Saint-Estèphe.

The Médoc peninsula has very little elevation, with peak altitudes reaching little more than 50 meters (165 feet) above sea level. However, there are gently sloping rolling hills throughout, locally named *croupes*, like those we visited farther south in Sauternes.

Rainfall is rather high in the area, which is to be expected for an ocean seafront, but it is somewhat tempered by the Landes forest to the west, which soaks up much of the rain. There is still a lot of hot, humid air that spills over the vineyard and acts as a temperature regulator: in the daytime, the water vapor filters out some of the solar radiation before it hits the ground, and during the nighttime it prevents some of the heat from escaping through the greenhouse effect.

North of Bordeaux, the first half of the Médoc peninsula is called Haut-Médoc, not because the altitude is that much higher or the wine is better, but because it is the upstream half. The wine does happen to be superior, though, and the first famous appellation we reach by driving half an hour out of the city is Margaux.

Margaux is a feminine-sounding name, and indeed the wine is said to be feminine as well, both delicate and voluptuous. When the wine is young, red berries make up the bouquet, but a few years in the cellar will bring out the complex aromas provided by Cabernet-Sauvignon.

The Margaux terroir is a good illustration of Médoc's division into a number of *croupes*. Looking at a topographic map, you can visualize the area as an "archipelago of rumps," separated from one another by small streambeds locally called *jalles*. When it rains hard, the water is quickly soaked up by the thick gravel beds—10 meters (about 30 feet) thick in places—and springs out into the *jalles*. As for the bedrock, which is rarely visible at the surface, it consists of lacustrine limestone called *calcaire de Plassac*, which dates to the Eocene period, over 40 million years ago.

The highest rump in Margaux, 25 meters (around 80 feet) above sea level and framed by two rivulets, bears an estate that is logically named Château du Tertre ("Château of the Knoll"). It is established on a Type 3 gravel bar, similar to the one we visited in the Sauternes area and aged approximately five hundred thousand years. It contains white and golden quartz, and blue and green volcanic pebbles rolled down from the Pyrenees Mountains. Be it the particular nature of this gravel bar or simply the winemaker's fancy, Château du Tertre is one of the rare Margaux that include a fair proportion of Cabernet Franc, a grape variety that brings strawberry, black currant, and blackberry into the wine's bouquet. Cabernet Franc accounts for as much as a third of the blend, on a par with Cabernet-Sauvignon and Merlot.

Besides Château du Tertre on this upper older terrace, most Grands Crus of Margaux are located on a central band of slightly lower hills, 20 meters (65 feet) above the estuary, that are classified as Type 4 terraces. Although the gravel is comparable to that found on the previous older level, here it is mixed with pockets of sand and clay. This combination gives a gravelly, brownish soil—technically called a brunisol—but some of the vineyards have so much gravel that there isn't any fine soil visible, as is the case at Château Margaux, the appellation's most famous estate (fig. 11.6).

To summarize, suave and delicate tannins and a fruity bouquet that becomes richly complex over time are the strong points of Margaux, and one need not seek a famous estate or pay a high price for it. At the south end of the district, Château des Graviers ("Château of Gravels") is a label that speaks for itself, at least as far as geology is concerned.

We visited the estate during harvesttime, when owner Christophe Dufourg-Landry had just started fermenting his Merlot (the Cabernet-Sauvignon needing a little more time to ripen in the field). At Château des Graviers, grape varieties are vinified separately and blended later—a custom in the Bordelais area, the other option consisting in harvesting and mixing all grapes in a single batch.

The other strategy followed by Château des Graviers is to let the grapes

Figure 11.6 At Château Margaux, cobbles deposited by the Garonne River both drain the terroir and store solar heat during the day, which help the grapes mature.

ferment naturally, without spiking the grape juice with commercial "designer" yeasts, although Christophe does experiment with them to advance research in the field. The final result is a fruity, aerial Margaux that definitely can claim a feminine touch.

PAUILLAC: CABERNET'S GRAND FINALE

Proceeding north past Margaux, the region of Haut-Médoc lines up one famous appellation after another, separated by small streams flowing down to the Gironde: Listrac, Moulis, Saint-Julien, Pauillac, and Saint-Estèphe.

Here we have the ultimate showdown between Cabernet and Merlot, with Cabernet reigning supreme over the gravel bars, and Merlot holding out on the lower slopes, where there is clay to be found.

One good example is provided by the adjacent townships of Listrac and Moulis, a couple of kilometers inland from the estuary, along route D 1215. In Listrac, the clay-rich slopes favor Merlot (60 percent), bringing lots of strength to the wine, which is tempered somewhat by the elegance and finesse of Cabernet-Sauvignon.

Slightly closer to the Gironde, we have Moulis (pronounced "Moo-lease"), a name that evokes windmills (*moulins*) that once spun their wings in the ocean breeze. Here the substrate is also clay-rich molasse, but it is

covered by large amounts of gravel that give Cabernet-Sauvignon the upper hand over Merlot. The Cabernet grape reaches 70 percent of the blend
where gravel hills are most prominent, and is responsible for the smooth
tannins that make Moulis such a harmonious, well-balanced wine.

Continuing our northward trek, we cross a swampy dip in the landscape, drained by a manmade channel (Jalle du Nord). We find ourselves
on a broad plateau, 4 by 4 kilometers (2½ by 2½ miles) in size, that looks
over the Gironde estuary and its riverside marshes. This is Saint-Julien
territory, an appellation that makes up for its relative flatness with a blanket of gravel 8 to 10 meters (26 to 33 feet) thick, similar to the Type 4 terrace that in Margaux carries most of the district's Grands Crus.

Thus, it should come as no surprise that the Saint-Julien appellation
favors Cabernet-Sauvignon, except in its westernmost part, where a layer
of sand mingles with the gravel and lowers the heat capacity of the ground
(smaller grain size), consequently favoring Merlot. Elsewhere, the gravel-
dominated terroir calls for 70 percent Cabernet-Sauvignon and 30 percent Merlot on average (with occasionally a few percent Cabernet Franc),
and results in a homogeneous, high-quality wine that varies little from
year to year.

One more rivulet to cross and we reach Haut-Médoc's nirvana, named
Pauillac. Its reputation goes back to the 1855 classification of the great
wines of Bordeaux, in which eighteen of the sixty nominated estates were
Pauillacs, including two of the four top-seeded Premiers Crus: Château
Latour and Château Lafite-Rothschild (the other two being Château
Margaux and Pessac-Léognan's Château Haut-Brion). This domination
was further bolstered in 1973 by the promotion of a third Pauillac to Premier Cru status: Château Mouton-Rothschild, located one kilometer
(sixth-tenths of a mile) south of its cousin Château Lafite (fig. 11.7).

The Pauillac township holds one of the highest concentrations of
gravel "rumps" in the entire Médoc peninsula, drained by multiple water-
ways running down to the estuary: the Juillac rivulet to the south, drawing the boundary with Saint-Julien; the Gaët channel in the center; and
the Jalle du Breuil to the north, drawing the line with Saint-Estèphe.

Again, the gravel bars are classified as Type 4 terraces and result from
the high discharge rate of the Garonne River a little less than a million
years ago. Gravel rumps are lined up along route D 2, which is truly
"Grand Cru Alley," since it passes through Saint-Julien, then Pauillac and
its two Rothschild estates, before entering Saint-Estèphe.

Figure 11.7 Château Lafite-Rothschild and its vineyard are located on a "rump" of gravel that slopes down to the Garonne River.

Cabernet-Sauvignon does so well on this gravel blanket that it accounts for up to 80 percent of the vines planted here. It brings to the wine both power—a strong alcohol content—and finesse, a balance that only a well-drained gravel terroir is capable of achieving.

At the northern end of Pauillac, where a major gravel rump slopes down to the Jalle du Breuil, Château Lafite-Rothschild stands out as one of the very best Pauillac and arguably the best red wines in the world. It relies on 80 percent to 95 percent of Cabernet-Sauvignon, depending on the year, with the balance being Merlot; sometimes a touch of Cabernet Franc or Petit Verdot is added. The blend is matured a year and a half in oak barrels, and the production is limited to twenty thousand cases of wine a year.

Lafite-Rothschild is famous for its bouquet of violet and almond, cedar, black currant, raspberry, and chocolate—with a whiff of black crayon and cigar-box tobacco—and for its exceptionally smooth tannins. Since imagination costs little, let us suggest to go with it some caramelized sweetbreads with truffles, or simply a savory beef rib cooked in the fireplace. For dessert, a second bottle of Château-Lafite calls for a chocolate cake. Don't forget to invite me.

CHAPTER TWELVE
The Rhône Valley

What better region with which to conclude our tour of French vineyards than one that encapsulates and summarizes the entire geological history of France? The Rhône valley is such a region. From Vienne in the north to Avignon in the south, it encompasses, over a distance of 200 kilometers (125 miles), just about every geological period we have encountered so far, in such prestigious terroirs as Côte-Rôtie and Hermitage, Tavel and Châteauneuf-du-Pape, Gigondas and Beaumes-de-Venise.

Driven toward the Mediterranean by a north–south-trending fault system, the Rhône is the unifying force running through all these vineyards. This mighty river flowing out of the Alps has carved steep slopes along its course and funneled winds and air masses to shape its own climate, bringing together continental, Mediterranean, and oceanic components.

The Rhône valley is also a cultural passageway in more ways than one. Both human culture and viticulture spread north and south along its banks: the Greeks and the Romans sailed up from the Mediterranean, and Helvetian tribes came down from their Alpine highlands. Roman merchants in particular settled along the Rhône decades before Julius Caesar and his legions conquered and unified the Gallic provinces (controlling the lucrative commerce of wine was apparently one of Rome's motivations for invading the land, according to some historians).

As for the invasion of vine, it had started as early as the second century BC. The ruins of winemaking domaines dating to this period were discovered near the city of Vienne, on the grounds of what is now the prestigious appellation Côte-Rôtie. The area became the first winemaking center along the river, exploiting grape varieties brought in from Switzerland and central Europe. Consequently, the city earned the well-deserved name of Vienne-la-Vineuse ("Vienne-the-winemaking-one").

Imported vine stock and crossbreeding experiments gave birth to a range of grape varieties that did especially well in the sunny Rhône climate, namely Syrah, Grenache, Mourvèdre, Cinsault, Roussanne, Marsanne, and Viognier. These grapes also colonized Languedoc and Provence, closer to the Mediterranean coast, but most give their best results in Côtes-du-Rhône.

Côtes-du-Rhône wines also owe their reputation to a twist of fate: the Avignon Papacy (1309–1378), when the seat of the pope was moved from Rome to southern France. Those were glorious years for Rhône vineyards, namely under the spiritual and very down-to-earth guidance of Pope John XXII, who had an old castle upstream of Avignon transformed into his summer residence. Around this Châteauneuf-du-Pape ("Newcastle of the Pope") a very successful vineyard was planted on steep, cobbly slopes. Centuries later, the vineyard would become one of the most prestigious appellations in the world. Côtes-du-Rhône wines have been on the map ever since, so it is only fair to end our tour of French terroirs by paying a visit to their most spectacular vineyards (fig. 12.1).

THE MICA SCHISTS OF CÔTE-RÔTIE

There is good reason to begin our tour at the northernmost Côtes-du-Rhône terroir, just outside Vienne. Besides being the oldest vine-growing area in the Rhône valley, producing today the sublime Cru of Côte-Rôtie, it boasts the oldest geological strata in the Côtes-du-Rhône region: the 300-million-year-old rocks that were once at the heart of the mighty Hercynian Mountains that ran across France and western Europe. As they also do in the Beaujolais region to the north, granites, gneisses, and mica schists from this Carboniferous period provide excellent soil and substrate to the vines.

The vineyard of Côte-Rôtie occupies the right bank of the Rhône, opposite the city of Vienne, and surrounds the town of Ampuis. The Renard rivulet that crosses the township divides the terroir into northern and southern wings, named respectively Côte Brune ("Brown Hill") and Côte Blonde ("Blond Hill") on account of their contrasting rock types. In the

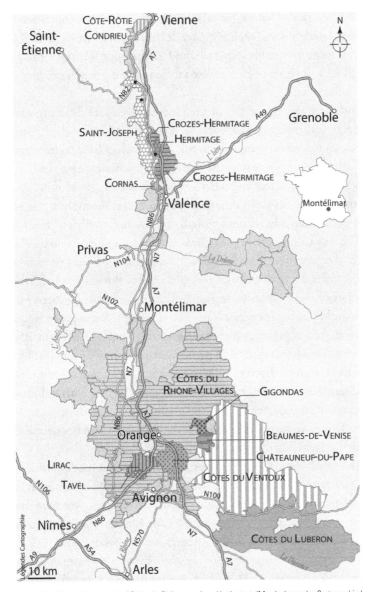

Figure 12.1 The main vineyards of Côtes-du-Rhône mentioned in the text. (Map by Legendes Cartographie.)

former, the rock breaks down into thin platelets and yields a mineral-rich soil, loaded with sparkling mica. This Côte Brune is landscaped into wide terraces, bound by walls of schist in order to keep erosion in check and prevent landslides.

Downstream of Ampuis, another type of metamorphic rock outcrops, named *leucogneiss* ("white gneiss"). Much paler in color, it contains mostly light-colored minerals such as quartz, feldspar, and white mica. This Côte Blonde yields a crumbly, clay-rich sand, so the slope is much more prone to failure and needs to be stabilized by landscaping into narrower, more frequent terraces.

Although the two wings of the vineyard are quite different when it comes to soil, their exposure for both is to the southeast—the Rhône follows a southwest-striking fault at this level—which is ideal for warming and drying the vines early in the day. Another advantage provided by the structural layout is that the valley segment is at an angle with the general north–south strike of the valley, a kink that breaks the momentum of the cold north wind, but still lets enough of it through to dry the vine leaves.

The very name Côte-Rôtie ("Roasted Coast") calls to mind the slope's hot and dry environment, where rain is scarce and the vine roots need to seek water at depth through clay-filled fractures. But above all, this terroir is famous for the exclusive and successful partnership—secured by two thousand years of experience—between it and the Syrah grape.

Syrah rules supreme over the entire northern half of Côtes-du-Rhône, where it does wonderfully well in the moderately hot climate (faring better here than in the hotter southern half) and crystalline substrate. Syrah produces deep-colored, tannic-rich, aromatic wines, with a dominant bouquet of black currant. The end result is particularly spectacular at Côte-Rôtie, where the two wings of the terroir—Côte Brune and Côte Blonde—yield two different versions of the wine.

The schist-based Côte Brune has deeper soil and more clay, which give the wine fruitier aromas, along with a stronger tannic structure. It is the more "masculine" of the pair, and its bouquet evolves over time toward forest undergrowth. The gneiss-based Côte Blonde is sandier, providing the wine with an elegant, flowery bouquet and a richer body.

Leaving Côte-Rôtie, we follow the Rhône downstream, looping around the rocky spur of Condrieu before striking a straight line to the south, with the vineyards clinging to the east-facing right bank and a newcomer challenging Syrah for a place in the sun. While Syrah here still produces excellent red wines, which go by the name of Saint-Joseph, the newcomer is Viognier, a grape that produces an exceptional white wine, golden yellow in color, giving off a bouquet of apricot and violet. The main Viognier appellation is Condrieu, named for the northernmost village that grows the variety. However, there is also a small enclave of Viognier that received a separate appellation—Château-Grillet. In this case, the grape is vinified in oak barrels and yields a wine famous both for its rarity (3½ hectares [8¾ acres] of vine; ten thousand bottles a year) and for its exquisite bouquet of apricot, honey, and truffle, with orange and tangerine joining in on the palate.

After this brief inroad by a white grape, Syrah and its red appellation Saint-Joseph reclaim full monopoly over the Rhône valley and its crystalline bedrock, which grades from gneiss and mica-schist upstream to granular granite downstream, altering into a coarse sand mixed with clay that also needs to be fixed in place by stone walls. This terroir brings out the elegant, velvety qualities of Saint-Joseph red wine, characterized by a bouquet of black currant, pear, licorice, and sometimes those subtle notes of forest undergrowth that we also find in Côte-Rôtie.

All the terroirs mentioned so far are located on the Rhône's right bank—the western Massif Central side, which has the steepest slopes and the best exposures. But as in all rules, there are exceptions. The one vineyard on the left bank, facing the city of Tournon and the vines of Saint-Joseph, is well worth mentioning, since it boasts two prestigious appellations: Hermitage and Crozes-Hermitage (see box).

BEAUMES-DE-VENISE AND THE TRIASSIC

We leave the ancient gneiss- and granite-based terroirs of the upper Rhône valley that date to the tectonic creation of France and proceed down the river, reaching the sedimentary relics of the great seas that invaded the country during the Triassic, Jurassic, and Cretaceous periods. Our destination is the southern Côtes-du-Rhône vineyards that stretch from Montélimar to Avignon and encompass such varied appellations as Valréas and Cairanne, Vacqueyras and Gigondas, Beaumes-de-Venise and Ventoux.

Beaumes-de-Venise in particular plays a pivotal role in the geological framework of the Côtes-du-Rhône landscape. The vineyard lies at the foot of a spectacular limestone massif tilted on edge—Les Dentelles de Montmirail—and spreads across the beveled layers of all three periods of the Mesozoic: Triassic, Jurassic, and Cretaceous.

The story of vine growing at Beaumes-de-Venise goes back to the classical era. Pliny the Elder (AD 23–79) already mentions the town's re-

---

HERMITAGE AND CROZES-HERMITAGE

The city of Tain-l'Hermitage, on the Rhône's left bank, is tucked inside a meander of the river, and surrounded north and south by a vineyard covering 1,400 hectares (3,500 acres): Crozes-Hermitage. A steep granitic hill rises above the city, and its small, 135-hectare vineyard is entitled to its own appellation, with the shorter name and even higher reputation of Hermitage.

The reputation of Hermitage goes back to the Renaissance. It was celebrated for its deep ruby color, "dark as ink"; its strong alcohol content; and its exceptional aroma. It still ranks today among the very best French wines in view of its strong tannins, great aging potential (10 to 20 years), and the complexity of aromas that time brings out, ranging from raspberry and blackberry to smoky and peppery flavors, with notes of leather, forest undergrowth, and musky game. There is also a small production of golden Hermitage white, made from Marsanne and Roussanne grapes, that has a distinctly floral bouquet of lime tree, iris, and narcissus.

The presence of a granitic hill on the eastern bank of the Rhône, belonging to the western Massif Central, is somewhat of an oddity. Leaning westward because of the regional slope set up by the Alpine uplift, the river abuts against the Massif Central and follows it south. At the level of Hermitage, however, the Rhône carved its way around a knob of granite, separating it from the massif and isolating it on the river's left bank.

This hill of granite bears the greater part of the Hermitage vineyard, and its wines need the most time to mature and blossom. The rest of the vineyard is established on a steep alluvial terrace of cobbles, east of the hill. Wines from this zone need fewer years to reach their prime and are fine and aromatic. At higher elevations, there is also a layer of windblown loess, brought in during the recent Ice Ages, that provides clay and powdery limestone to the soil and nourishes the Roussanne and Marsanne grapes of the superb Hermitage whites.

Beyond the granitic hill of Hermitage and its eastern extension of cobbles, the wider appellation of Crozes-Hermitage spreads across eleven townships, but lacks the steep slopes, good drainage, and ideal exposure of its towering neighbor. The wines follow a similar terroir-driven trend: the granitic soil in the northern part of the appellation yields well-structured wines that age well, whereas the gravel terraces of the southern part yield aromatic wines that should be drunk young.

---

nowned grape in his encyclopedia *Historia Naturae*: "The Muscat grape has been grown for a long time in Beaumes and its wine is remarkable." Beaumes's reputation hadn't faded a bit by the Middle Ages, since we know that in 1248 King Louis the Saint stopped there to stock up on wine on his way to the Seventh Crusade, and Pope Clement V (circa 1264–1314), having left Rome to settle in the south of France, also relied on Beaumes-de-Venise for wine.

Muscat was the main grape variety at the time, and is still used today to produce one of the most popular sweet wines in France. It boasts a rich, golden color and a peachy bouquet—a perfect match for chocolate cakes and desserts. Beaumes-de-Venise also produces an excellent dry red from Grenache and Syrah grapes.

The terroir of Beaumes-de-Venise is especially complex. The site underwent a major upheaval during the rise of the Alps, as it happens to straddle a deep fracture zone—the Nîmes fault—that runs westward across Provence from the Alps to the Pyrenees.

Before the fault system sprang into action, the area was a lowland flooded by a shallow sea during the Triassic period, over 200 million years ago. As was the case with Alsace to the north (see chapter 3), the shallow lagoons evaporated under sunny climes, leaving behind thick layers of salt, such as gypsum. Later on, during the Jurassic, the region became a deep basin that collected clay-rich marl on its bottom—sediments that are now dark gray and called *terres noires* ("black earth") by the winegrowers. Finally, as the sea level dropped somewhat during the Cretaceous, a period that began 145 million years ago, coral reefs grew on the site, and their light-toned, calcareous debris is referred to as *terres blanches* ("white earth").

This layering would have remained horizontal and orderly, with the Triassic gypsum at the bottom and the Cretaceous limestone on top, had the Alpine uplift not disturbed the pile about 20 million years ago. The old Nîmes fault sprang into action, and the displacement was spectacular. The soft layer of Triassic gypsum deformed and flowed up the fault, acting like a lift gate to raise the thick sequence of Jurassic and Cretaceous limestone above it, and tilt that on edge. The sequence now points skyward, sculpted by erosion into sharp crests called the Dentelles de Montmirail (*dentelles* meaning "lace," so fine-looking is the crest line) (fig. 12.2).

The cooperative winery of Beaumes-de-Venise, Balma Venitia, is proud of the variety of strata outcropping in its district. It markets three different red wines that correspond to the three geological periods represented in the upturned strata: ochre-colored terroir for the Triassic, "gray earth" for the Jurassic, and "white earth" for the Cretaceous.

The Terres du Trias cuvée is a rich wine that takes full advantage of the old claystone and gypsum squeezed up along the fault. The soil is iron rich, hence the ochre color, with streaks of white and violet, and has just enough porosity to drain the rainwater and grant the vines only the minimal amount of water and nutrients necessary, making them work hard for their living and concentrate aromatic molecules. Grenache and Syrah

Figure 12.2 Many estates of Beaumes-de-Venise AOC highlight the limestone needles of the Dentelles de Montmirail on their wine labels.

are blended together and extract from the gypsiferous soil a bouquet of black berries and Mediterranean scrubland, and silky tannins that reach a harmonious balance. This is the kind of wine to drink with game.

Terre des Farisiens is the Beaumes-de-Venise cuvée raised on the gray, calcareous terroir of the Jurassic, which is found on the southeastern flank of the Dentelles de Montmirail. The marine sediment, compacted and sliced up into thin layers, is easily penetrated by the rootstock. On slopes facing southeast, the wine has a more earthy and roasted bouquet. It calls for a lamb tajine cooked with olives—a good match, since olive trees are plentiful on the Montmirail slopes.

The third and last cuvée of this geological triptych, Terres de Bel Air represents the "white earth" of the Cretaceous period: the youngest of the lot. This terroir is found in the center of the sedimentary "pocket bread" folded along the Nîmes fault around the picturesque, perched village of La Roque-Alric. Over a kilometer (six-tenths of a mile) thick on average, but emerging at the surface only in places, these Cretaceous strata are cultivated in terraces. The resulting wine has a fresh, fruity bite, Italian-style, that goes very well with a beef carpaccio appetizer or with osso bucco (braised veal shanks).

GIGONDAS AND THE JURASSIC

Beaumes-de-Venise is not the only appellation to benefit from the tectonic uplift that squeezed and folded sediments along the Nîmes fault. On the northwestern flank of the Dentelles de Montmirail lies the 1,200-hectare (3,000-acre) vineyard of Gigondas, set on gravelly slopes reaching down to the Ouvèze River valley—the western border of the appellation (fig. 12.3).

The red wines of Gigondas are powerful and spicy, and benefit from the great interplay between the Grenache grape and the local terroir. Grenache makes up 80 percent of the blend, complemented by a touch of tannic Syrah and aromatic Mourvèdre. Gigondas's rich bouquet brings together lots of black: black pepper, black cherries, blackberry, and black currant. Its best cuvées rival Châteauneuf-du-Pape, but at a more affordable price.

Gigondas is ideally located upslope of the Ouvèze valley at the convergence of two crest lines that dominate the village: La Grande Montagne des Dentelles to the north and Les Dentelles Sarrasines to the east. As is also the case at Beaumes-de-Venise, it is generally the softer layers of the

Figure 12.3 The Gigondas AOC vineyard occupies the lower slopes of the Dentelles de Montmirail, up to the tree line.

sedimentary sequence that host the vineyard, particularly its southeastern slopes.

When the grapes come from these marl and limestone slopes, Jurassic or Cretaceous in age, Gigondas wine is known to be rich and powerful, with the best aging potential. There is also a second type of terroir for the wine on the lower slopes south of the village, where gravel and cobble terraces were draped over the bedrock by the Ouvèze River during the high runoffs of the recent Ice Ages. This dual terroir constitutes the major division that runs through the broad aromatic palette of Gigondas wine. To explore the range of wines in detail, drop by the cellar of the winemakers association* and sample the hundred or so estates and cuvées that are sold there . . . before booking a room for the night.

TAVEL AND THE CRETACEOUS

After the Triassic, Jurassic, and Cretaceous terroir of Beaumes-de-Venise and Gigondas, we cross the Rhône valley to explore further the Cretaceous period and the terroir laid down by the marine invasions that swept through France. We head for Tavel on the river's right bank, 15 kilometers (9 miles) west of Avignon, climbing up the slope toward a plateau that leads inland to the Cévennes region: an arid landscape covered with scrubland, olive trees, and of course grapevines.

The winemaking tradition here goes back to antiquity, but it was during the Middle Ages that Tavel got its break, seducing the papal court of Avignon. Barrels of Tavel were even shipped to Rome, where it was heralded as the "wine of the Gods." Tavel was also transported up the Rhône valley to Paris and Versailles, and caught the attention of Dutch wine traders. Moreover, it became a favorite of Louis XIV—him again!—and later of nineteenth-century writers Honoré de Balzac and Alphonse Daudet.

Tavel is a rosé wine, a color and wine type that it brought to perfection. Many wine critics rank it as one of the best if not the best rosé in France, and perhaps the world. It was famous in the past for its bronze color, deep aroma, and good aging potential. Today, oenologists point out its bouquet of red berries leaning toward raspberry, its spicy touch, and its

---

*Caveau du Gigondas, place de la Mairie, 84190 Gigondas. Open every day from 10:00 a.m. to 12:00 p.m. and from 2:00 to 6:00 p.m. (7:00 p.m. in the summer). Phone: 04 90 65 82 29. Website: http://www.gigondas-vin.com/decouvrir/le_caveau_du_gigondas.php.

Figure 12.4 In its upper, western part, the Tavel AOC vineyard is planted on Cretaceous limestone, which was broken up into cobbles by freeze-thaw action during the recent Ice ages.

overall smoothness. It goes well with cold cuts, veal stew, lamb, and fish in sauce.

Tavel relies mostly on Grenache, with variable proportions of Cinsault, Mourvèdre, and Syrah. The maceration of grape juice and skins lasts up to two days, loading the wine with color and tannin.

The Tavel terroir is divided into two wings. The western wing is a lacustrine limestone plateau with a central dip carved by the Malaven stream, followed by road D 4 (fig. 12.4). The limestone layers, 130 million years old, belong to the Early Cretaceous period. Originally coral reefs, they are hard and yield a special type of terroir, because the recent Ice Ages, through water infiltration, freezing, and expansion, broke up the rock into white, angular rubble. These platelets, locally named *lauses*, were used to build stone walls in the villages and fields. Such a stony terroir reflects the sun's rays onto the vines, heats up the soil, and brings vivacity to the wine.

The eastern wing of Tavel is the slope leading down to the Rhône. This is a very different terroir, covered with fist-sized cobbles that the river dumped as a high-perched terrace, roughly 2 million years ago, before digging its course down to its present level, 30 meters (100 feet) lower. Grenache and Cinsault grown on this wing's terraces yield an even deeper-colored wine, highly structured.

Before leaving the Rhône's right bank, the nearby village of Lirac, just upstream from Tavel, is worth a visit. Here, too, the Rhône has spread its cobbles. Known at first for its rosé in trying to compete with Tavel, Lirac has now turned to red. Its reputation is fast growing, built on Grenache, Syrah, and Mourvèdre, along with a fruity contribution from Cinsault. Red Lirac is a strong, tannic, well-structured wine known to age well, joining Gigondas as a serious challenger to Châteauneuf-du-Pape.

CHÂTEAUNEUF-DU-PAPE

Facing Tavel and Listrac on the opposite bank of the Rhône, Châteauneuf-du-Pape is home to one of the most famous wines in the world. This wine is recognizable by its squat bottle bearing the city's emblem embossed on the glass: the papal tiara and the two keys of Saint Peter, leading to Earth and to Paradise.

Fifteen kilometers (9 miles) upstream of Avignon, the hill of Châteauneuf-du-Pape benefited greatly from the move of the papal seat to Avignon during the fourteenth century, when a civil war was brewing in Rome and French King Philip IV the Fair was trying to gain control over the Catholic Church. Elected pope in 1316 at the age of seventy-two, John XXII settled in that riverside city, which possessed an Episcopal palace—this was upgraded to a papal palace (Palais des Papes)—and a bridge over the Rhône River that connected Provence and Italy to Languedoc and Spain.

A short and fragile-looking man—the cardinals had elected him in their belief that he would last only a few more years—John XXII surprised everyone by ruling close to twenty years (he died in 1334 at the age of ninety). He worked hard to consolidate the papacy and develop the Avignon region.

A native of Cahors in southwestern France—his real name was Jacques Duèze—the pontiff was fond of wine, and so had the best winemakers from his home region come to Avignon to plant and supervise new vineyards. These included plantings on the hill north of the papal city, where the Episcopal castle mentioned earlier was renovated to become his summer residence: Châteauneuf-du-Pape ("Newcastle of the Pope"). John XXII also purchased the Côtes-du-Rhône vineyard of Valréas—

and, while he was at it, the whole town with it—and had Muscat grapes planted at Malaucène, halfway between the Dentelles de Montmirail and Mont Ventoux.

MONT VENTOUX

Referred to sometimes as *le géant de Provence* ("the Provence giant"), Mont Ventoux is an isolated mountain that stands apart from the Alpes-de-Haute-Provence limestone range to the east, the Baronnies range to the north, and the Dentelles de Montmirail to the northwest. Peaking at 1,912 meters (6,373 feet) above sea level, Mont Ventoux's conical profile has it often mistaken for a volcano, an impression strengthened by a white cap that looks like snow or fallen ash. It is, however, a limestone mountain, and its cap is simply the naked rock, light toned and bearing no vegetation.

The uplift of the Pyrenees mountain range at the close of the Cretaceous period pushed limestone blocks to the north into a great fold that was later sculpted into Mont Ventoux. The mountain became an island when the Mediterranean Sea spread into the Rhône valley and flooded the Provence basins around it, during the Miocene epoch 20 million years ago.

The mountain has its own AOC appellation: Ventoux (formerly called Côtes-du-Ventoux). Its vineyard is located mostly at the foot of its southern slope. It yields principally red wine (85 percent of the appellation), but also rosé (15 percent) and some white (1 percent). Vine lots are scattered on all favorable spots around the mountain, notably on the slopes of the nearby Malaucène basin rim, in the direction of Beaumes-de-Venise, and in the Carpentras amphitheater, which shares its limestone terroir between vineyards for wine, vineyards for table grapes, and fruit tree orchards.

The limestone of Mont Ventoux is of Jurassic age, but more recent limestone underlies the vineyards; it was deposited during the Mediterranean Sea advance of the Miocene epoch, when the mountain was an island. Terraces of even more recent gravel blanket the Carpentras basin.

Ventoux red wines are as varied as their terroir can be, with quite a range of winemaking practices. They range from fruity light reds with a bouquet of red berries to powerful, deeply colored cuvées that age well and develop a bouquet of prune and cherry liquor over time.

At Châteauneuf-du-Pape today, the vineyard has expanded to cover the entire hill, which slopes steeply down to the Rhône River to the west; the Ouvèze River and its confluents to the east; the plains of Orange to the north; and the alluvial terrace of Sorgues and Avignon to the south. As many as thirteen different grape varieties are authorized in its AOC area, proof of a great deal of experimentation and vine-growing success over the centuries. In practice, though, Grenache is by far the dominant variety (70 percent of the vineyard), followed by Mourvèdre, Syrah, and Cinsault, the nine other varieties playing a very minor role.

Châteauneuf-du-Pape red wine is famous for its deep-garnet to purple color; strong alcohol content (up to 14 proof); surprising smoothness; great aging potential (several decades for the best vintages); and a rich bouquet of ripe fruit, spices, and roasted coffee, evolving over time toward anise and licorice, leather, musk, and truffle. Châteauneuf-du-Pape also puts out an excellent white (7 percent of the total production) that has a bouquet of honeysuckle and narcissus and is full and long on the palate.

The hill of Châteauneuf-du-Pape is protected from the Rhône's erosive power by a pedestal of hard Cretaceous limestone, but it is the layers on top that make up the substrate of the terroir: strata of sand and sandstone laid down by marine incursions during the Miocene epoch, 20 million years ago. At the time, the Mediterranean Sea flooded a moat lining the nascent Alpine mountain range, penetrating as far north as present-day Switzerland. The trench filled with sand, calcareous seashells, and debris shed from the growing mountains. This clay-rich molasse outcrops halfway up the hill of Châteauneuf-du-Pape, and is visible in the embankments of the road that leads down to Bédarrides on the southern slope.

The marine invasion and sedimentation ended abruptly 6 million years ago, due to extraordinary circumstances. Plate tectonic movements between Europe and Africa ended up pinching and closing the Straits of Gibraltar. With no more water supplied by the Atlantic Ocean, the Mediterranean Sea experienced a negative balance, losing more water by evaporation than it could gain from river discharge into its basin. This we know for a fact, because thick beds of salt, concentrated by the evaporation process, are found in Mediterranean sediments of that age. According to models, evaporation lowered the sea level in the basin by half a meter (1½ feet) per year, taking less than two thousand years to drop it by a full kilometer (six-tenths of a mile) — and it might have dropped lower yet.

This Mediterranean crisis had profound repercussions throughout the region. The Rhône gained momentum to keep up with the drop in sea level. Turned into a raging whitewater river, it dug itself a deep canyon, several hundred meters below its normal course. This high-energy situation, with the Straits of Gibraltar closed, lasted nearly five hundred thousand years.

At the end of this interval, tectonic readjustments opened up the straits once again. Atlantic waters rushed back into the Mediterranean basin, filling it to the brim and drowning the Rhône canyon. The marine waters swept up the valley as far north as Lyon, transforming the gorge into a peaceful fjord that progressively filled up with silt and sediments washed down from the Alps. Blue marl from this filling episode outcrop today along the Rhône valley, in particular east of Crozes-Hermitage and around Châteauneuf-du-Pape, but it is of little interest to vine growing. This comeback of the Mediterranean Sea during the Pliocene epoch, less than 5 million years ago, funneled water around the hill of Châteauneuf-du-Pape. It became a surf-pounded island, where large quartz cobbles piled up, brought down from the Alps by strong currents.

When the sea level returned to normal, slowly retreating from the hill, cobbles spread down the slopes, with some clay mixed in. Now left high and dry, this thick blanket of tawny-colored cobbles is a spectacular sight. It wraps around the entire hill, flush with the vineyards (fig. 12.5).

The cobbles readily soak up the sun's heat during the day and radiate it back to their surroundings at night, helping the grapes mature. For tapping the necessary moisture, vine roots can rely on the clay matrix that fills the gaps between the cobbles. For additional moisture and minerals, they also reach the sandy molasse of the Miocene sea at a few meters' depth, beneath the cobbles.

This overview of the geological history of France, so well summarized in the Côtes-du-Rhône, would not be complete without noting the Ice Ages that span the last 2 million years of the Earth's history. At Châteauneuf-du-

Figure 12.5 The terroir of Châteauneuf-du-Pape is famous for its large cobbles, laid down on the slopes of what once was a surf-beaten island. In the background are the town and also the medieval castle that was the summer residence of the Avignon popes.

Pape, as well as at Tavel on the opposite bank, we saw how the Rhône piled up terraces of cobble and gravel carried down from the Alps. Elsewhere, around the cities of Sorgues and Orange, for instance, the terraces are sensitive to droughts, because there is little clay trapped between the stones, but vines still do rather well on this terroir, and grapes mature quickly.

Here ends our tour of French terroir, begun 500 million years ago in the schists of Anjou and reaching the dawn of civilization in the Ice Age terraces of the Rhône valley.

But the story is far from over. Global warming is changing vine-growing conditions at a rapid pace, affecting temperatures and precipitations so that the climatic, geographical boundaries of each grape variety are fast shifting. New territories, lacking vines today, might inherit favorable conditions and become the great winemaking terroirs of tomorrow. Geologists and entrepreneurs are already testing the soil and potential of many areas of northern France in this respect.

This is a whole new story that will take place in the near future. To witness it, we will need to be patient and live a long life, following the sage advice of François Rabelais—the Renaissance man from Chinon: "Drink always and you shall never die."

CHÂTEAUNEUF-DU-PAPE AOC

Region:	Southern Côtes-du-Rhône
Wine type:	red (mostly), white
Grape variety:	red: Grenache (70%), Mourvèdre, Syrah, Cinsault, etc. (13 varieties authorized)
	white: Clairette, Bourboulenc, Roussanne
Area of vineyard:	3,150 hectares (7,875 acres)
Crus:	famous climates and estates (such as Le Vieux Télégraphe)
Nature of soil:	sand, quartzite cobbles, clay
Nature of bedrock:	sandstone, limestone
Age of bedrock:	Miocene (20 million years ago)
Aging potential:	5 to 20 years (reds); 3 to 5 years (whites)
Serving temperature:	16–18°C (61–64°F) (reds); 10–11°C (50–52°F) (whites)
To be served with:	lamb, beef, hare, duck, wild boar, venison

GLOSSARY

ammonites: Group of marine mollusks belonging to the **cephalopod** class, with a spiral shell divided into chambers; flourished from the **Triassic** until the End **Cretaceous** periods.

amphibole: Dark-colored mineral of volcanic and **metamorphic** rocks, rich in iron and magnesium.

amphibolite: **Metamorphic** rock made up mostly of **amphibole** minerals.

andesite: Volcanic rock, relatively rich in silica (57% to 63%), containing feldspar, pyroxene, and **amphibole** minerals.

ankylosaurs: Group of herbivorous quadruped dinosaurs armored with bony plates and spikes (osteoderms); lived during the **Jurassic** and **Cretaceous** periods.

anticline: An arch of rock strata in which the layers bend downward from the top in opposite directions, with the oldest rocks in the central core.

AOC: Appellation d'Origine Controlée, a French certification granted to certain wines, cheeses, and other agricultural products from a distinct geographical area.

appellation: A certification of origin or quality delivered by a ruling body or government bureau.

Aquitanian: Division of the **Miocene** epoch, ranging from 23 million to 20.4 million years before present.

argillaceous: Containing clay.

Bajocian: Division of the Jurassic period, ranging from 172 million to 168 million years before present.

basin: A closed topographic low area, usually filled with sediments.

Bathonian: Division of the Jurassic period, ranging from 168 million to 165 million years before present.

bauxite: Aluminum ore, mixed with clays and iron oxides.

bedform: A depositional feature in sediments that is indicative of flow (water current or wind).

belemnite: Extinct group of marine cephalopods, related to the modern cuttlefish.

bivalves: Class of mollusks living in a shell consisting of two hinged parts, closed by ligaments.

breccia: Rock consisting of rounded or angular fragments assembled together in a finer matrix, which can be produced by fluvial processes, landslides, volcanic explosions, or asteroid impacts.

caillotte: In the Sancerre region, local name of calcareous rocks broken into fist-sized platelets and cobbles by freeze-thaw action.

calanque: Steep-walled inlet or cove, usually formed in limestone, found along the Mediterranean coast.

calco-alcaline: Chemical suite of magmatic rocks rich in calcium and alkaline metals (potassium, sodium).

Cambrian: Geological period ranging from 540 million to 488 million years before present.

Campanian: Division of the Upper Cretaceous period, ranging from 83.5 million to 70.6 million years before present.

Carboniferous: Geological period ranging from 359 million to 299 million years before present.

Cenomanian: Division of the Upper Cretaceous period, ranging from 99.6 million to 93.5 million years before present.

Cenozoic: Geological era regrouping the Tertiary (a term no longer officially used) and the Quaternary eras, from 65.5 million years ago to the present.

cephalopods: Class of marine mollusks, including extant squid and octopi and extinct **ammonites** and **belemnites**, characterized by a prominent head surrounded by tentacles.

chalk: Form of **limestone** (marine **sedimentary** rock), generally white, light, and porous, composed mainly of fine, calcareous debris of plankton and algae **tests.**

clay: Mineral type structured in sheets of tiny crystals rich in aluminum and silicon. Also, the rock composed of clay minerals.

climate (*climat*): In viticulture, a small area characterized by its soil and microclimate. Synonymous with **Cru.**

clos: Small area of a vineyard (*climat* or **Cru**) enclosed by a stone wall.

crémant: Sparkling wine, named originally because its lower carbon dioxide pressure gave it a creamy rather than a fizzy mouth-feel.

Cretaceous: Geological period ranging from 144 million to 65.5 million years before present.

crinoids: Class of marine animals belonging to the echinoderm phylum; includes the sea lily, which attached itself to the sea bottom by a stemlike tube from which emerged a bouquet of filtering arms, giving the animal the appearance of a flower.

croupe: Low hill, named for the French word for "rump."

Cru (and Grand Cru, Premier Cru): Synonymous with *climat*, a small area of a vineyard defined by its specific location (terrain and microclimate). By implication, *Cru* also defines the wine produced in such an area, and **appellations** of Grand Cru, Premier Cru, and so on refer to their classification according to quality.

cuesta: A ridge formed by tilted layers of **sedimentary** rock, truncated by erosion into a steep slope (ex.: cuesta of Île-de-France in the Champagne region).

cuvée: Wine of a specific blend or batch.

dalle Nacrée: Literally "mother-of-pearl slab" in French, it is the name given to a shiny **Jurassic limestone** found in Burgundy.

desiccation crack: Open fissure in a **clay**-rich ground that dries up and contracts.

Devonian: Geological period ranging from 416 million to 359 million years before present.

diorite: A bluish to dark-gray magmatic rock relatively rich in silica (52 to 63%).

dromaeosaurs: Family of bird-hipped theropod dinosaurs, feathered and carnivorous (ex.: *Velociraptor*), that were widespread during the Upper **Cretaceous** period.

Eocene: Geological epoch ranging from 55.8 million to 33.9 million years before present.

estuary: River mouth characterized by brackish water.

fermentation: Transformation of an organic substance by bacterial enzymes, including the transformation of sugar into alcohol.

flexure zone: Area where the Earth's crust is bent by the piling up of heavy sediments or other stress associated with tectonic motions.

foehn: Warm and dry alpine wind, and by extension any mountain-based dry wind.

foehn effect: The drying of soil and vegetation by a **foehn**.

fossé. See **rift zone**.

gneiss: Metamorphic rock, often of granitic composition, with a banded texture (alternating dark and light bands) caused by deformation under heat and pressure.

graben. See **rift zone**.

granite: Silica-rich magmatic rock, crystallized at depth and composed mainly of large, interlocking feldspar and quartz crystals, with some minor **amphibole** and **mica**.

griotte: In the Sancerre region, small *caillotte* (see this word).

gypsum: Sulfate mineral ($CaSO_4.2H_2O$), tabular and translucent, precipitated by the evaporation of salt-saturated lakes and seas.

hadrosaurs: Family of ornithopod (bird-hipped) herbivorous dinosaurs with a duck-billed snout; flourished during the Upper Cretaceous period.

hardpan: Dense layer of soil that restricts the passage of water.

horst: Positive elevation flanking the down-dropped part of a rift zone (the graben), and often referred to as the "shoulder" of the rift.

isostasy: Gravitational equilibrium of the upper layers of the Earth, governed by density differences and buoyancy forces.

Jurassic: Geological period ranging from 201.6 million to 144.5 million years before present.

Kimmeridgian: Division of the Upper Jurassic period, ranging from 156 million to 151 million years before present.

lacustrine: Relating to a lake.

lignite: Brown coal of low carbon content (25 to 35 percent) and high moisture.

limestone: Marine or lacustrine sedimentary rock composed of calcium carbonate, mostly of biogenic origin (compacted shells and plankton tests).

loess: Wind-blown, silt-sized sediment, often derived from glacial environments.

Maastrichtian: Division of the Upper Cretaceous period, ranging from 70.6 million to 65.5 million years before present.

maceration: Step of the winemaking process in which raisins are placed in a vat, and the fermenting juice acquires color, tannin, and aromatic molecules from its contact with the grape skins.

magma: Mineral mass that has undergone partial or total melting. Essentially liquid, it often also contains crystals in suspension and dissolved gases.

malolactic (fermentation): **Fermentation** step in which tart-tasting malic acid is transformed by anaerobic bacteria into softer-tasting lactic acid.

marl: **Sedimentary** rock of low permeability, composed of a fine mix of **limestone** and **clay**.

massif: Section of the Earth's crust bounded by faults, often shaped into a mountain range.

metamorphic (rock): Rock that has experienced high temperature and pressure at depth in the Earth's crust; characterized by a special suite of altered minerals and an often banded texture.

mica: Mineral structured around silicon and aluminum oxides, rich in potassium, sodium, calcium, iron and magnesium. It splits cleanly into flakes.

mica schist: **Metamorphic** rock composed of sheets of **mica**-rich minerals and often showing high cleavage (splits easily into its constituent sheets).

mildew: Plant disease caused by the growth of small parasitic fungi that can affect vine leaves and fruit; it is favored by rain and high humidity.

Miocene: Geological epoch ranging from 23 million to 5.3 million years before present.

molasse: Clay-rich **sedimentary** rock formed in underwater **basins** along mountain fronts; it provides debris material through erosion.

monzonite: Light-colored **granite**, with abundant feldspar and little quartz.

mudrock: Fine-grained **sedimentary** rock, such as siltstone, claystone, and **shale**.

must: Pressed grape juice, containing grape skins and seeds, that undergoes **maceration** and **fermentation** to become wine.

noble rot: Benevolent form of gray fungus—Botrytis cinerea—that affects certain grapes, lowering their water content and increasing sugar and aromatic molecules.

Oligocene: Geological epoch ranging from 33.9 million to 23 million years before present.

Ordovician: Geological period ranging from 488 million to 444 million years before present.

ornithopods: Dinosaur group of families characterized by a birdlike hip joint and three-toed feet.

Oxfordian: Division of the Upper Jurassic period, ranging from 161 million to 156 million years before present.

Paleocene: Geological epoch ranging from 65.5 million to 55.8 million years before present.

phylloxera: Parasitic sap-sucking insect that infects vine roots with a poisonous secretion.

piedmont: Foothills marking the transition from mountain range to low plains.

Pliocene: Geological epoch ranging from 5.3 million to 1.8 million years before present.

Portlandian (also called Tithonian): Division of the Upper Jurassic ranging from 151 million to 145.5 million years before present.

pyroclastic: Containing broken-up particles of magmatic (volcanic) origin.

regolith: Layer of loose material (dust or soil) covering solid rock.

rhyolite: Viscous, silica-rich lava that can take on many colors (green, yellow, red, black in its glassy obsidian form).

rift zone: Down-dropped **basin** along sets of parallel faults, caused by the thinning and extension of the Earth's crust (synonyms: **graben**, *fossé*).

Rognacian: Division of the Upper **Cretaceous** period, defined in the south of France, ranging from 70 million to 65 million years before present and roughly corresponding to the internationally recognized **Maastrichtian** division.

rootstock: Vine variety resistant to diseases and parasites (**phylloxera** in particular), upon which other varieties of grapes can be grafted.

sandstone: Sedimentary rock formed by the accumulation and consolidation of sand grains, under or above water.

sauropods: Group of quadruped and lizard-hipped herbivorous dinosaurs, characterized by their great size, long tail, and long neck.

schist: Finely layered **metamorphic** rock, often derived from **clays** and muds, that often splits easily along its planar layers (ex.: **mica schist** and slate).

scoria: Pebble-sized volcanic ejecta, porous and permeable.

scree: Broken rock fragments at the base of a steep hill or cliff; also known as talus deposits.

sedimentary (rock): Type of rock formed by the piling up, under or above water, of organic or inorganic mineral debris or precipitated salts.

shale: Fine-grained **sedimentary** rock, formed of **clay** minerals and other silt-sized particles.

Silurian: Geological period ranging from 444 million to 416 million years before present.

Sparnacian: Division of the Lower **Eocene** epoch, ranging from 55.8 million to 53 million years before present; the term is used in France and is named for the Champagne town of Épernay. It is included in the internationally recognized Ypresian division (55.8 million to 48.6 million years before present).

spilite: Ancient altered basalt, light to dark green in color, containing alteration minerals enriched in sodium.

stegosaurs: Group of bird-hipped herbivorous dinosaurs characterized by a small head, a long neck, and a double row of bony plates on their back and tail; flourished during the **Jurassic** and Early **Cretaceous** periods.

subduction: Motion of one tectonic plate under another at a convergent boundary, often accompanied by earthquakes and volcanic activity.

suture zone: Area where separate geographical areas or crustal blocks are spliced together along faults.

syncline: A trough of rock strata in which the layers curve downward, with the youngest rocks in the central core.

talus: See **scree**.

test: Inner shell of plankton and other primitive organisms, built of silica or calcium carbonate.

theropods: Suborder of lizard-hipped, bipedal, and mostly carnivorous dinosaurs.

titanosaurs: Subgroup of large **sauropod** dinosaurs (quadruped, lizard-hipped, and herbivorous) with a small, flattened head; flourished during the Upper **Cretaceous** period.

trachyandesite: Siliceous volcanic lava, rich in feldspar, light to dark gray in color, that often constitutes thick flows, plugs, and domes.

Triassic: Geological period ranging from 251 million to 201.6 million years before present.

tuffeau: White, sparkling **chalk** of the Loire valley containing small amounts of quartz, opal, and **mica**; it is quarried and used as building stone for many monuments and châteaux.

Turonian: Division of the Upper **Cretaceous**, ranging from 93.5 million to 89.3 million years before present.

varietal: Wine made from a single type of grape variety, and by extension often used to name the grape variety itself.

vein: Mineral body, often emplaced as hot **magma** or as a hydrothermal fluid, that fills a fissure zone and solidifies. Some rock veins contain economically interesting mineral ores.

BIBLIOGRAPHY

PUBLICATIONS

Bernardin, E., and P. Le Hong. 2010. *Crus Classés du Médoc, le long de la route des Châteaux.* Bordeaux: Éditions Sud-Ouest.

Bousquet, J.-C., and M. Vianey-Liaud. 2001. *Dinosaures et autres reptiles du Languedoc.* Montpellier: Les Presses du Languedoc.

Dupin, T. 2008. *Le Lutétien moyen de Fleury-la-Rivière.* Paris: Société Amicale des Géologues Amateurs.

France, B., ed. 2008. *Grand atlas des vignobles de France.* Paris: Solar.

Frankel, C. 1996. *The End of the Dinosaurs: Chicxulub Crater and Mass Extinctions.* Cambridge: Cambridge University Press.

———. 2007. *Terre de France: Une histoire de 500 millions d'années.* Paris: Le Seuil.

Joseph, R. 2005. *French Wine.* New York: DK Publishing.

Pomerol, C., ed. 1980. *Geology of France.* Paris: Masson, collection Guides géologiques régionaux.

———, ed. 2003. *Wines and Winelands of France.* Paris: Éditions du BRGM.

Tastet, J.-P., P. Becheler, and J.-C Faugères. 2003. *Géologie et typicité des vins de Bordeaux, Livre excursion, 9ème Congrès Français de Sédimentologie.* Paris: ASF.

Wever, P. de, et al. 2009. "Géologie et vin." *Géologia* 97, 4–15.

WEBSITES

French vineyard information and news (*English*): http://www.winetourisminfrance.com/an/

Oenology lessons in France (*French*): http://www.oenologie.fr/

Oenology tours in France (*English*): http://www.guideduvignoble.fr/english/index.html

Vineyard-visiting itineraries (*French*): http://www.hachette-vins.com/tourisme-vin/route-vins/

ALSACE

Association of independent winemakers of Alsace (*French*): http://www.alsace-du-vin
.com/
Etienne Loew estate (*English*): http://www.domaineloew.fr/vin-alsace/?lang=en
Touristic route of wines in Alsace (*English*): http://www.alsace-route-des-vins.com
/NewVersion/index.cfm/Language/En.cfm
Geological sites of Alsace, University of Strasbourg (*French*): http://www.lithotheque
.site.ac-strasbourg.fr/
Weinbach estate (*English*): http://www.domaineweinbach.com/en/
Wines and vineyards of Alsace (*English*): http://www.vinsalsace.com/en

ANJOU (LOIRE VALLEY)

Closel estate (Château des Vaults), Savennières (*English*): http://www.savennieres
-closel.com/-English-.html
Coulée-de-Serrant estate, Savennières (*English*): http://www.coulee-de-serrant.com
/en/index-en.html
Guide of Savennières and Loire Valley wines (*English*): http://www.richardkelley
.co.uk/
Roche-aux-Moines estate, Savennières (*English*): http://www.domaine-aux-moines
.com/en/menu1E.html

BEAUJOLAIS

Association of Beaujolais winemakers (*English*): http://www.beaujolais-wines.com/EN/
Association of Fleurie AOC winemakers (*English*): http://www.cavefleurie.com/en/
Association of Morgon AOC winemakers (*French*): http://www.morgon.fr
Association of Moulin-à-Vent AOC winemakers (*French*): http://www.moulin-a-vent
.net/
Beaujolais Geological Museum, Saint-Jean-des-Vignes (*French*): http://www.espace
-pierres-folles.com/
Brouilly and Côte-de-Brouilly AOC (*French*): http://www.espace-des-brouilly.com
/index.php
Georges Duboeuf Wine Park "Hameau du Vin" (*English*): http://www.hameauduvin
.com/#/en/
Les Roches bleues estate, Côte de Brouilly (*French*): http://www.lesrochesbleues.fr/
Official site of Beaujolais wines (*French*, *English*): http://www.beaujolais.com/
Daniel Rampon estate, Morgon and Fleurie (*French*): http://www.domaine-daniel
-rampon.com/

BORDEAUX

Château Cheval Blanc, Saint-Émilion (*English*): http://www.chateau-cheval-blanc
.com/
Château-Figeac, Saint-Émilion (*English*): http://www.chateau-figeac.com/index_en
.php

Château des Graviers, Margaux (e-mail): chateau.des.graviers@orange.fr
Château Gueyrosse, Saint-Émilion (*English*): http://gueyrosse.free.fr/
Château Lafite-Rothschild and other estates (*English*): http://www.lafite.com/eng
Château Mangot, Saint-Émilion (*English*): http://www.chateaumangot.fr/
Château Margaux (*English*): http://www.chateau-margaux.com/
Château Mille-Secousses estate, Côtes de Bourg (*French*): http://www.mille-secousses
   .com/
Château Suduiraut, Sauternes (*English*): http://www.suduiraut.com/
Château du Tertre, Margaux (*English*): http://www.chateaudutertre.fr/uk/
Château d'Yquem, Sauternes (*English*): http://www.yquem.fr/
Cru d'Arche-Pugneau estate, Sauternes (*French*): http://arche.pugneau.free.fr/
Moulis AOC, Haut Médoc (*French*): http://www.moulis.com/
Vineyards of Bordelais area (*English*): http://www.vignobledebordeaux.fr/index
   .php?lang=en

BURGUNDY

Burgundy building and decoration limestone (*English*): http://www.pierre
   -bourgogne.fr/en
Burgundy wines (*French*): http://www.divine-comedie.com/
Domaine Denis (Corton, Pernand-Vergelesses) (*English*): http://www.domaine
   -denis.com/
Dubreuil-Fontaine estate, Corton (*English*): http://www.dubreuil-fontaine.com
   /en/index.html
Pierre Marey and *fils* estate, Corton and Pernand-Vergelesses (*French*): http://www
   .topfrenchwines.com/domaine-pierremarey/
Official site of Burgundy wines (*English*): http://www.burgundy-wines.fr/

CENTRAL LOIRE VALLEY

Association of Saumur-Champigny winemakers (*English*): http://www.producteurs
   -de-saumur-champigny.fr/
Bernard Baudry estate, Chinon (*English*): http://www.chinon.com/vignoble
   /Bernard-Baudry/ENG_default.aspx
Chinon AOC (*French*): http://www.chinon.com/
Delanoue Frères estate, Bourgueil and Saint-Nicolas-de-Bourgueil (*English*): http://
   www.domaine-de-la-noiraie.com
Domaine de la Perruche estate, Saumur-Champigny (*French*): http://www.domaine
   delaperruche.com/
Saint-Nicolas-de-Bourgueil AOC (*French*): http://www.stnicolasdebourgueil.com/
Saumur-Champigny, Château de Parnay (*French*): http://www.chateaudeparnay.fr/
Saumur-Champigny, Antoine Cristal's Clos des murs (*French*): http://www
   .closdesmurs-cristal.com
Troglodyte dwellings of Anjou (*French*): http://hades.troglodyte.free.fr/anjou.htm

CHAMPAGNE

La Cave aux coquillages (Fleury-la-Rivière): Champagne and fossil digs (*French*): http://www.lacaveauxcoquillages.fr/

Champagne winemakers association (*English*): http://www.champagnesdevignerons .com/

Côte des Bars: Champagne and Rosé des Riceys (*English*): http://www.champagne -les-riceys.com/GB/

Official site of Great Brands and Champagne Houses (*English*): http://www.maisons -champagne.com/en/

Tarlant estate, Œuilly (*English*): http://www.tarlant.com/en/champagne.htm

LANGUEDOC

Savary de Beauregard estate (*English*): http://savarydebeauregard.com/

Coteaux du Languedoc wines (*English*): http://www.coteaux-languedoc.com/en

Dinosaur Museum, Espéraza (*French*): http://www.dinosauria.org/

Dinosaur Park and Museum, Mèze (*English*): http://www.musee-parc-dinosaures .com/anglais/

Faugères AOC wines (*French*): http://www.faugeres.com/

Limoux AOC wines & sparkling wines (*French*): http://www.limoux-aoc.com/

Saint-Chinian AOC wines (*English*): http://www.saint-chinian.com/en.php

Joseph Salasar estate, Limoux sparkling wines (*French*): http://www.salasar.fr/

MÂCONNAIS/POUILLY-FUISSÉ

Association of Pouilly-Fuissé AOC winemakers (*French*): http://www.pouilly-fuisse .net/

PROVENCE

Bandol AOC wines (*English*): http://www.vinsdebandol.com/

Cassis AOC wines (*English*): http://www.maisondesvinscassis.com/en/

Château de la Bégude, Côtes de Provence, Rousset (*English*): http://www.chateau delabegude.com/website/chateau-de-la-begude4/index4.asp

Château de Fontblanche, Cassis (*French*): http://www.guidevins.com/pages/gvp /cassis/fontblanche.html

Clos Magdeleine estate, Cassis (*English*): http://www.clossaintemagdeleine.fr/

Côtes de Provence Sainte-Victoire AOC (*French*): http://www.vins-sainte-victoire .com/

Domaine de Saint-Ser estate, Côtes de Provence Sainte-Victoire (*English*): http:// www.saint-ser.com/

Ferme Blanche estate, Cassis (e-mail): fermeblanche@wanadoo.fr

Mas de Cadenet estate, Côtes de Provence Sainte-Victoire (*English*): http://www .masdecadenet.fr/international/homepage.htm

Wines of Auriol and surroundings (*French*): http://www.vigneronsdugarlaban.com/fr

RHÔNE VALLEY

Beaumes-de-Venise AOC winemakers (*French*): http://www.cavebalmavenitia.com/
Beaumes-de-Venise wines (*French*): http://www.beaumesdevenise-aoc.fr/
Châteauneuf-du-Pape AOC (*French, English*): http://www.chateauneuf.com/english/
Wines of the Rhône valley (*English*): http://www.vins-rhone.com/en
Wines of Tavel AOC (*French*): http://www.cavedetavel.com/
Wines and terroir of Gigondas AOC (*English*) http://www.gigondas-vin.com/

UPPER LOIRE VALLEY AND SANCERRE

Henri Bourgeois estate (Sancerre) at Chavignol (*English*): http://www.henribourgeois
.com/
Centre-Loire wines (*French*): http://www.vins-centre-loire.com/vignobles/
Domaine de Bel-Air estate, Pouilly-sur-Loire (*English*): http://www.bel-air-pouilly
.com/index.php?LANGUE=1
P. Girardin, Atlas of landscapes of the Cher Department (*French*): http://www.cher
.pref.gouv.fr/atlas-cher/
Vincent Grall estate, Sancerre (*English*): http://www.grall-vigneron-sancerre.com/
Menetou-Salon AOC wines (*English*): http://www.menetou-salon.com/
Sancerre winemakers association (*English*): http://www.maison-des-sancerre.com
/en/.

GEOGRAPHICAL AND
WINE NAMES INDEX

Page numbers in italics indicate captions.

GRAPE VARIETIES INDEX

PERSONAL NAMES INDEX

Printed in the USA
CPSIA information can be obtained
at www.ICGtesting.com
LVHW100957281223
767564LV00004B/512

9 780226 816722